D0918569

INFRARED, CORRELATION, AND FOURIER TRANSFORM SPECTROSCOPY

COMPUTERS IN
CHEMISTRY AND INSTRUMENTATION

edited by

James S. Mattson | Harry B. Mark, Jr. | Hubert C. MacDonald, Jr.

Volume 1: Computer Fundamentals for Chemists

Volume 2: Electrochemistry: Calculations, Simulation, and Instrumentation

Volume 3: Spectroscopy and Kinetics

Volume 4: Computer-Assisted Instruction in Chemistry (in two parts).
Part A: General Approach. Part B: Applications

Volume 5: Laboratory Systems and Spectroscopy

Volume 6: Computers in Polymer Sciences

Volume 7: Infrared, Correlation, and Fourier Transform Spectroscopy

INFRARED, CORRELATION, AND FOURIER TRANSFORM SPECTROSCOPY

EDITED BY

James S. Mattson

National Oceanic and Atmospheric Administration
Environmental Data Service
Center for Experiment Design and Data Analysis
Washington, D.C.

Harry B. Mark, Jr.

Department of Chemistry
University of Cincinnati
Cincinnati, Ohio

Hubert C. MacDonald, Jr.

Koppers Company, Inc.
Research Department
Monroeville, Pennsylvania

MARCEL DEKKER, INC. New York and Basel

Library of Congress Cataloging in Publication Data
Main entry under title:

Infrared, correlation, and fourier transform spectroscopy.

(Computers in chemistry and instrumentation ; v. 7)
Includes bibliographical references and indexes.
1. Infra-red spectrometry--Data processing.
2. Fourier transform spectroscopy--Data processing.
3. Minicomputers. I. Mattson, James S. II. Mark, Harry B. III. MacDonald, Hubert C.
QD96.I5I54 543'.085 77-9460
ISBN 0-8247-6369-6

MARCEL DEKKER, INC.
270 Madison Avenue, New York, New York 10016

Current printing (last digit):
10 9 8 7 6 5 4 3 2 1

PRINTED IN THE UNITED STATES OF AMERICA

INTRODUCTION TO THE SERIES

In the past decade, computer technology and design (both analog and digital) and the development of low cost linear and digital "integrated circuitry" have advanced at an almost unbelievable rate. Thus, computers and quantitative electronic circuitry are now readily available to chemists, physicists, and other scientific groups interested in instrument design. To quote a recent statement of a colleague, "the computer and integrated circuitry are revolutionizing measurement and instrumentation in science." In general, the chemist is just beginning to realize and understand the potential of computer applications to chemical research and quantitative measurement. The basic applications are in the areas of data acquisition and reduction, simulation, and instrumentation (on-line data processing and experimental control in and/or optimization in real time).

At present, a serious time lag exists between the development of electronic computer technology and the practice or application in the physical sciences. Thus, this series aims to bridge this communication gap by presenting comprehensive and instructive chapters on various aspects of the field written by outstanding researchers. By this means, the experience and expertise of these scientists are made available for study and discussion.

It is intended that these volumes will contain articles covering a wide variety of topics written for the nonspecialist but still retaining a scholarly level of treatment. As the series was conceived it was hoped that each volume (with the exception of Volume 1 which is an introductory discussion of basic principles and applications) would be devoted to one subject; for example, electrochemistry, spectroscopy, on-line analytical service systems. This format will be followed wherever possible. It soon became evident, however, that to delay publication of completed manuscripts while waiting to obtain a volume dealing with a single subject would be unfair to not only the authors but, more important, the intended audience. Thus, priority has been given to speed of publication lest the material become dated while awaiting publication. Therefore, some volumes will contain mixed topics.

The Editors

PREFACE

For years, while much of the spectroscopic instrumentation market was ignoring the inroads being made into the laboratory by minicomputers, and now by microprocessors, individual researchers were deeply involved in the mating of laboratory "workhorse" instruments and powerful laboratory computers. As Norman Jones states in Chap. 1 of this volume, his laboratory began the move to computer-aided infrared data handling in 1960. The major infrared instrument manufacturers did not introduce a computer-interfaced spectrophotometer to the commercial market until 1975, years after similar systems were in existence in several university and industrial laboratories around the country.

Chapters 1 and 2 of this volume focus on infrared spectroscopic data-handling possibilities (Chap. 1) and laboratory data acquisition (Chap. 2). Both of these authors have used the same commercial infrared spectrophotometer as their data source in their laboratories. In Chap. 1, N. Jones of the National Research Council of Canada describes fifty computer programs developed in his laboratory over a period of fifteen years, from simple transmittance-absorbance conversion routines to complicated band deconvolution programs. In Chap. 2, J. Mattson and C. A. Smith describe an on-line minicomputer data system for the Perkin-Elmer model 180 infrared spectrophotometer in sufficient detail that the reader needs little else to be able to duplicate their system. The data system described by Mattson and Smith in Chap. 2 would make an ideal complement to the data reduction programs described by Jones in Chap. 1.

In Chap. 3, R. Wiens and H. Zwick of Barringer Research Ltd. share the innovative notions of correlation spectroscopy applied to measurement of gaseous pollutants in the atmosphere. They describe the whole range of correlation spectroscopic techniques from the dual-chambered gas-filled detector developed by Luft in 1938 to a satellite-borne, Fourier transform correlation spectrometer developed by Barringer Research to locate sources and sinks of CO in the earth's atmosphere. Correlation Fourier transform spectroscopy borrows from many fields of modern spectroscopy to provide a singularly powerful remote sensing device for single gaseous species, for example, SO_2, NO, CO. Wiens and Zwick provide the reader with enough of the theory of correlation spectroscopy, and descriptions of the requisite hardware and electronics, that individuals with only a rudimentary background in analytical spectroscopy will be able to understand their chapter. Furthermore, there is sufficient detail in the chapter that some may be tempted to go out and build their own instrument.

In Chap. 4, C. Foskett of Digilab delves into the rationale behind choosing a laboratory minicomputer, comparing the needs of a dispersive infrared spectrophotometer with a Fourier transform infrared interferometer. Foskett explains the requirements for a laboratory computer in terms of the rate of information to be transferred and the information content of each datum. In describing the criteria to be employed by a potential minicomputer user, Foskett uses a Data General NOVA as his "model" system. His discussion thus nicely complements Mattson and Smith's chapter, where the same family of minicomputers is employed in conjunction with a dispersive infrared spectrophotometer.

Washington, D. C.

James S. Mattson
Harry B. Mark, Jr.
Hubert C. MacDonald, Jr.

CONTRIBUTORS TO THIS VOLUME

CHARLES T. FOSKETT, Digilab, Inc., Cambridge, Massachusetts

R. NORMAN JONES, Division of Chemistry, National Research Council of Canada, Ottawa, Canada

JAMES S. MATTSON,* Rosenstiel School of Marine and Atmospheric Science, University of Miami, Miami, Florida

CARROLL A. SMITH, Rosenstiel School of Marine and Atmospheric Science, University of Miami, Miami, Florida

R. H. WIENS, Barringer Research, Ltd., Rexdale, Ontario, Canada

H. H. ZWICK,† Barringer Research, Ltd., Rexdale, Ontario, Canada

*Current affiliation: National Oceanic and Atmospheric Administration, Environmental Data Service, Center for Experiment Design and Data Analysis, Washington, D.C.

†Current affiliation: Canada Centre for Remote Sensing, Ottawa, Ontario, Canada

CONTENTS

INFRARED, CORRELATION, AND FOURIER TRANSFORM SPECTROSCOPY

Chapter 1

MODULAR COMPUTER PROGRAMS FOR INFRARED SPECTROPHOTOMETRY

R. Norman Jones

Division of Chemistry
National Research Council of Canada
Ottawa, Canada

I. INTRODUCTION

Our laboratory first used a computer to analyze infrared spectral data in 1960 when we developed a program to compute the second, third, and fourth truncated moments to quantify the shape and asymmetry of infrared absorption bands [1]. The spectrum was measured on a single-beam grating spectrometer and the chart was hand-digitized in terms of the wave number and the absorbance. The program was written in machine language for the IBM 1620 computer; a description was published in 1962 [2].

In 1961 we began to write programs in FORTRAN II, still working with data digitized from the recorder chart and punched on Hollerith cards. These programs were written in connection with the preparation of wave number calibration tables [3] and for the intensity calibration of infrared spectrophotometers with high-speed rotating sectors [4]; IBM 650 and 1620 computers were used.

Our experience with direct digital data logging began in 1963 when we acquired a Perkin-Elmer Model 421 spectrophotometer with direct encoding on paper tape. The organization of the set of programs that forms the basis of this chapter began at that time. The data logging system, as it was operating in 1967, has been described [5,6]. Parts of that system are still in use for the measurement of attenuated total reflection spectra, but the main work of the laboratory is now channeled through a Perkin-Elmer Model 180 spectrophotometer with magnetic tape encoding. Once in storage the data can be accessed from a time shared terminal in the laboratory with an on-line plotter adjacent to the terminal.

During the past decade most of our work has involved computer processing of digitally recorded spectral data and a wide range of programs has been written. In our own operations most of the programs are incorporated as subroutines in automated systems for spectrophotometer calibration, data reduction, and spectral analysis, but concomitant with the development of this integrated system a modular set of independent programs has been prepared. Twenty-two of these modular programs were published in 1968-1969 [7-9] and an additional 28 more recently developed programs will soon be available [10-13]. They are listed by title in an appendix to this chapter and may be obtained from our laboratory (see note 1).

In this chapter the functions and general structure of the programs are outlined. They are all written in FORTRAN IV and are designed for card input and card and printed output. Each program executes a specific type of spectrophotometric calculation and collectively they will perform most of the basic calculations of absorption spectrophotometry. A defined set of abscissal and ordinate scales is used and there are programs for scale and unit conversion. The FORTRAN terminology and the input and output card formats are standardized to facilitate the interfacing of the programs; in most cases the numerical sections of the card output from one program can be used directly as input to others.

Our laboratory deals mainly with the infrared spectra of condensed phase systems, but the use of the programs in other fields of vibrational and electronic spectroscopy is facilitated by Program II which provides for interchange between wave number scales in reciprocal centimeters and wavelength scales in micrometers and angstroms with retention of equal interval abscissal digitization.

It must be emphasized at the outset that in writing these programs the main intent has been to convey the algorithms in a readable form, and no claim is made for high computational efficiency. It is a truism that as computer programs are edited to increase their efficiency as means to communicate between man and machine, they become decreasingly effective as means to communicate between mind and mind through direct human intelligence. It was not our purpose in preparing these programs to create a series of computationally efficient "black boxes." Rather we wished to convey to chemists and spectroscopists some of our experience in formulating the underlying algorithms in computer readable language. One would anticipate that each user will wish to adapt the programs to operate with optimal efficiency within the framework of the configuration of his own computer system. This applies particularly to the very simple card image input and output formats which are capable of considerable compression, especially if the user is prepared to forego the generalized interfacing potential and link them directly as subroutines. Storage requirements, of particular importance to those working with minicomputers, can be reduced by more efficient overlay of the arrays and more sophisticated nesting of the do-loops.

II. ORGANIZATION OF THE RAW SPECTRAL DATA

The standard formats for the abscissal and ordinate scales used for card output in these programs are summarized in Table 1. The abscissal units of wavelength and wave number are the conventional ones in molecular spectroscopy except for $cm^{-1} \times 10$ (see note 2). This is a convenient unit for the manipulation of infrared wave number data with five-figure precision in fixed point arithmetic.

TABLE 1

Standardized Card Input and Output Formats

	Transmittance	Transmittance × 1000	Absorbance	Molar absorption coefficient	Absorption index
Cm^{-1}	6(F6.1,F6.3,1X)	7(F6.1,I4,1X)	6(F6.1,F6.3,1X)	5(F6.1,F7.2,1X)	5(F6.1,F7.4,1X)
$Cm^{-1} \times 10$	6(I5,F6.3,1X)	8(I5,I4,1X)	6(I5,F6.3,1X)	6(I5,F7.2,1X)	6(I5,F7.4,1X)
Micrometers	6(2F6.3,1X)	7(F6.3,I4,1X)	6(2F6.3,1X)	5(F6.3,F7.2,1X)	--
Angstroms	5(F8.1,F6.3,1X)	6(F8.1,I4,1X)	5(F8.1,F6.3,1X)	5(F8.1,F7.2,1X)	--

This precision is adequate for most measurements on condensed phase systems where round off to 0.1 cm^{-1} is admissible.

Mention should be made of the variety of ordinate units used in these programs. It is assumed that the measurements are made on the conventional type of double-beam spectrophotometer.

The *transmittance* T_ν is given by

$$T_\nu = \left(\frac{P}{P_0}\right)_\nu \tag{1}$$

where $P_{0(\nu)}$ is the radiant flux of the reference beam and P_ν the radiant flux transmitted by the material in the sample beam with the spectrometer set at the wave number ν. T is used only occasionally, and it is preferable to use T × 1000 which can be manipulated in fixed-point arithmetic with photometric precision commensurate with the capabilities of contemporary spectrophotometers [4-6,14].

The *absorbance*

$$A_\nu = \log\left(\frac{P_0}{P}\right)_\nu$$

$$= \log T_\nu^{-1} \tag{2}$$

Both T and A are dimensionless (see note 3).

The *molar absorption coefficient*

$$\varepsilon_\nu = \frac{A_\nu}{cd} \tag{3}$$

where c is the concentration (mmol cm^{-3}) and d the path length (cm). The dimensions of ε are $mmol^{-1}$ cm^2. The term *molar absorptivity* is also used for ε. Where dispersion theory is involved it is convenient to express the absorption intensity in terms of the *absorption index*

$$k_\nu = \frac{\ln(P_0/P)_\nu}{4\pi\nu d}$$

$$\sim \frac{2.30258\ A_\nu}{4\pi\nu d} \tag{4}$$

The dimensionless quantity k_ν is the imaginary component of the complex refractive index $\hat{n}_\nu$ which, following the IUPAC sign conventions, is written

$$\hat{n} = n + ik \tag{5}$$

where n is the simple refractive index (cf. Sec. V.B). These sign conventions also require that the electric vector of the radiation be written

$$E = E_o \exp[2\pi i\omega t] \tag{6}$$

where ω is the frequency (rad sec^{-1}).

A. Units and Scale Changes

Programs I, II, and XXVII provide for interchange among the various unit systems listed in Table 1; they differ in their complexity and sophistication. The simplest is Program I, which will accept data in the unit systems i-iv of Table 2 and provide output in the systems i-vi. It will not change the wave number encoding interval. Program XXVII is an updated version of Program I with facility to output in the systems i-viii.

TABLE 2

Unit Interchanges Available in Programs I and XXVII

Unit system	Wave number unit	Intensity unit
i	cm^{-1}	T × 1000
ii	cm^{-1} × 10	T × 1000
iii	cm^{-1}	A
iv	cm^{-1} × 10	A
v	cm^{-1}	ε
vi	cm^{-1} × 10	ε
vii	cm^{-1}	k
viii	cm^{-1} × 10	k

Program II is more flexible. The input provides for the separate designation of the abscissal unit and the ordinate unit. The output can be specified in any combination of these except for the trivial μm ⇌ Å conversion. An interpolation subroutine using a five point Lagrange function [16] is included to allow for equal interval wave number output from equal interval wavelength input and vice versa. This interpolation subroutine (TERPOL) can also be used to alter the number and spacing of the equal interval wave number data points without change of the wave number unit.

When used in the wave number ⇌ wavelength mode, data can be handled over the range 1000.0 to 999,999.9 Å to deal with most problems likely to be encountered in connection with visible and ultraviolet spectra.

B. Manual Digitization of Data

Although digitally recording spectrophotometers are becoming increasingly common, the fact remains that most chemical spectroscopy is dependent on chart plotted output. Many of these programs can be used to good effect to analyze data generated on simple spectrophotometers using chart recorders if the data are first transcribed from the chart to punched cards.

Program XXIII is modified from one written by T. Bulmer and H. F. Shurvell [17] who have been using these programs in this manner. The program reduces the tedium and potential inaccuracies in preparing such manually punched input card data. The initial experimental record must be in the form of a chart on a linear abscissal scale. The scale may be wave number with either the low or high values to the left. Alternatively the curve may be plotted on graph paper in any linear abscissal scale, e.g., centimeters or inches (the "chart readings"). The ordinate scale should be transmittance, %-transmission or other scale easily converted to T × 1000. The input parameters to the program are the beginning and terminating wave numbers in cm^{-1} and the corresponding beginning and ending chart readings in floating point arithmetic. The desired wave number encoding interval must also be specified. The program generates a print out with the wave numbers tabulated at the required intervals, a second column lists the corresponding chart readings, and a third column is left blank. The intensity ordinates are then written into this column by hand in units of T × 1000 by inspection or measurement from the chart.

The program also prepares a card deck in the standard 8(I5,I4,1X) format for input to the program collection. The card deck, prepunched in the wave number columns, is stacked in the card reader and the T × 1000 ordinates are punched in manually from the printed table. This operation is facilitated by using an appropriately punched drum-card to skip the $cm^{-1} \times 10$ fields. Bulmer and Shurvell also describe how Program IV (cf. Sec. III.A) can be used to check the completed card deck for ordinate punching errors.

C. Sequencing Checks on Card Decks

In data transfer by punched cards misplacement of a card is a potential source of error. In many of these programs a subroutine SORT is introduced to monitor this. Obviously this subroutine should be removed if the program is modified for tape input or direct input transfer from computer files. SORT subroutines are written separately for each program so that the programs remain self-contained; there are also minor variations between the different SORT subroutines to take account of specific requirements of the individual programs.

SORT first establishes that the starting wave number and the wave number encoding intervals are correct. An error here will terminate the program with an appropriate message. This message also serves as a diagnostic of an omitted parameter card in the input deck. The successive abscissal values are next checked against the wave number interval. If a mismatch occurs which corresponds to an exact multiple of the wave number interval, it is assumed that a card has been wrongly sequenced. The faulty data point is transferred to a separate storage register, as are the succeeding points until the correct sequence is reestablished. In the commonly used 8(I5,I4,1X) format this should occur at the ninth succeeding point if only one card is misplaced; for the misplacement of n cards the correct sequence will be established after 8n + 1 false abscissal values. SORT will store a total of 160 misplaced data points. If the sequencing errors exceed this, the program is terminated with an error message.

If the number of sequence errors is less than 160, the data are rearranged in the correct sequence and the computation proceeds. A table listing the corrected input data is printed. If an abscissal value not corresponding to an integral multiple of the wave number interval is detected the computation is terminated and an error message printed. If no error is found the computation proceeds with the message "Wave number sequence is correct."

D. Reorganization and Concatenation of Card Decks

With card or simulated card input the need can arise to generate new decks by concatenation. This can present difficulties if the junction is to occur where the terminal data field of the first deck is not the last data field on a card, or the first data field of the second deck is not the first data field on a card.

Program XXIV permits the generation of concatenated card decks provided only that the concatenation involves the last card of the first deck and the first card of the second deck. The program will deal with multiple decks in a similar fashion. It operates only in the standard 8(I5,I4,1X) format for data encoded as $cm^{-1} \times 10/T \times 1000$. The concatenated data are checked for continuity before the new deck is punched.

E. Linearization and Normalization of the Baseline

Situations arise where it is required to linearize the baseline of a spectrum defined by two points on a curve, such as A and B of Fig. 1. These points are commonly tangential to the curve but not necessarily so.

Operating in one mode, Program XXV will recompute the ordinates in terms of a baseline AC parallel to the abscissal axis. In an alternative mode it will also displace the curve to align the base on

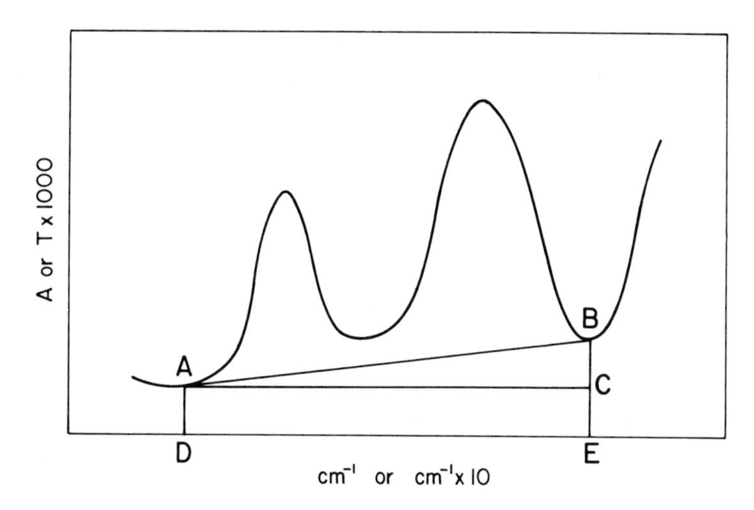

FIG. 1. Program XXV. Linearization and normalization of the baseline.

the axis (DE of Fig. 1). The input ordinates can be transmission (T × 1000) or absorbance but in either case the computations are made in absorbance units. The output may be in units of cm^{-1} × 10/T × 1000 or cm^{-1}/A. This algorithm is also incorporated into Programs VIII and XXI.

III. BASIC ANALYSIS OF THE SPECTRAL CURVES

In this section several programs are described that are concerned with the processing of the spectral data after they have been converted to a "clean" set of coordinates and freed from gross experimental imperfections and discontinuities. In processing spectral data in our laboratory the experimental measurements, as directly digitized on paper or magnetic tape, are first submitted to an editing and normalizing process.. This takes account of "catastrophic" errors in intensity due to line surges or other abnormal extraneous noise in the generation and digitization of the raw data and to dropped abscissal points. It also applies corrections for wave number and intensity calibration errors. We have not incorporated these programs into the modular set since they are complex and not well adapted to card input format. Some of the individual sections of the data editing process are discussed in Sec. IV, but it is more convenient to deal first with the smoothing and differentiation of the spectra, the location of band maxima and minima, band area summation, and related matters.

A. Smoothing and Differentiation

There is an extensive mathematical literature concerning the smoothing of digitized data. In most spectroscopic work the favored method to improve the signal-to-noise ratio is to convolve the data with an appropriate smoothing function. The basic theory will be found in the standard textbooks of statistics [18] and its application to chemical spectroscopy was developed by Savitzky and Golay [19], whose article provides the basis of most smoothing algorithms currently used by spectroscopists. Their treatment was essentially empirical and a more quantitative analysis of the smoothing problems of infrared spectrophotometry has been made by Porchet and Günthard [20].

Our practical experience with infrared spectra of condensed phase systems encoded at 0.5 cm^{-1} intervals (see note 4) suggested that effective suppression of high-frequency noise with minimal distortion of the true band shape could be achieved with a five-point quadratic smoothing function of the form

$$y_{s(i)} = \frac{1}{35}\left[-3(y_{i-2} + y_{i+2}) + 12(y_{i-1} + y_{i+1}) + 17y_i\right] \tag{7}$$

where $y_{s(i)}$ is the smoothed value of the ith ordinate [21,22]. Subsequently, we found a slight improvement with a nine-point quartic function

$$y_{s(i)} = \frac{1}{429}\Big[15(y_{i-4} + y_{i+4}) - 55(y_{i-3} + y_{i+3}) + 30(y_{i-2} + y_{i+2}) + 135(y_{i-1} + y_{i+1}) + 179y_i\Big] \tag{8}$$

Similar types of functions will generate the first and second derivative curves. For the first derivative we use a nine-point quadratic function,

$$\left(\frac{dy}{dx}\right)_i = \frac{1}{60(\delta\nu)}\left[4\left(y_{i+4} - y_{i-4}\right) + 3\left(y_{i+3} - y_{i-3}\right) + 2\left(y_{i+2} - y_{i-2}\right) + \left(y_{i+1} - y_{i-1}\right)\right] \tag{9}$$

where $\delta\nu$ is the encoding interval.

For the second derivative a 13-point quadratic function is used

$$\left(\frac{d^2y}{dx^2}\right)_i = \frac{1}{1001(\delta\nu)^2}\left[22\left(y_{i-6} + y_{i+6}\right) + 11\left(y_{i-5} + y_{i+5}\right) + 2\left(y_{i-4} + y_{i+4}\right) - 5\left(y_{i-3} + y_{i+3}\right) - 10\left(y_{i-2} + y_{i+2}\right) - 13\left(y_{i-1} + y_{i+1}\right) - 14y_i\right] \tag{10}$$

In Program IV these smoothing and differentiating algorithms are applied. The program accepts input data as cm^{-1}/T, $cm^{-1} \times 10/T$, $cm^{-1}/T \times 1000$, $cm^{-1} \times 10/T \times 1000$, cm^{-1}/A, $cm^{-1} \times 10/A$. The use of the cm^{-1}/T and $cm^{-1} \times 10/T$ scales are advantageous in generating the second derivative curves since they are less susceptible to round off error and oscillation; most sections of this program operate in double precision.

In the smoothing modes, the quantity $\Delta y = y_{smoothed} - y_{unsmoothed}$ is evaluated for each ordinate, and the quantities $\Sigma\ \Delta y/n$ and $\Sigma\ \Delta y^2/n^2$ are computed, where n is the number of ordinates; they are listed together with Δy_{max} and ν_{max}. This provides some simple statistical information about the effectiveness of the smoothing process; Δy_{max} and ν_{max} identify the data point which has suffered the greatest change in ordinate intensity. In the smoothing mode the program can be operated cyclically to smooth repeatedly until Δy_{max} is reduced below a preset value or a specified number of smoothing cycles has been performed.

In the differentiating modes the operator has the option of accepting the output directly or first submitting the derivative ordinates to a preset number of smoothing cycles. It must be born in mind that on each differentiation or smoothing cycle a number of ordinates are dropped at each end of the wave number range.

If the ordinates are entered as absorbance the differentiation will be performed on the absorbance ordinates but all smoothing is carried out on T × 1000 ordinates in floating point arithmetic. The final smoothed curve will be reconverted to absorbance if that form of output is requested. The change from absorbance to transmission is done because the signal-to-noise ratio of experimental data will normally be largely independent of T but will vary logarithmically with A. The smoothing will therefore be more uniform on the T scale.

B. Location of Maxima and Minima

Program III scans the spectrum to locate the positions of the extrema. In one mode it will print a table of the positions and intensities of the absorption maxima; in a second mode it will similarly list the minima and in a third mode it will list both maxima and minima. Program XXX is an updated version which has an option which makes it more efficient in dealing with the sharp narrow peaks of gas phase spectra. Neither program will locate inflections.

These programs are modifications of the PEAKFIND program developed by Savitzky [19,23]. The basic step is to evaluate the first derivative at successive encoding points across the spectrum and locate where a change of sign occurs. A sign change of $dA/d\nu$ from + to - identifies an absorption peak and from - to + an absorption valley (see note 5).

Both Program III and Program XXX accept input in cm^{-1}/A and $cm^{-1} \times 10/T \times 1000$ units. The ordinates are first normalized with respect to the baseline and the zero transmission line. The slope of the curve is then determined at each ordinate using a 25-element first derivative convoluting function in Program III and a choice of this or a 9-element convoluting function in Program XXX. When an ordinate for which $dT/d\nu$ is zero is located (or when the sign between two consecutive values of $dT/d\nu$ changes), the direction of the sign change is checked and, if appropriate, the program switches to a branch which analyzes the peak region. First, the ordinate difference from the preceding extremum is evaluated. If this is less than a preset value the point is ignored and the examination of the curve proceeds. This mechanism provides a noise rejection gate. A common operating value for this corresponds to a 3% peak-to-peak signal-to-noise ratio. If the zero slope point is accepted as a genuine peak the program next passes a least-squares cubic convoluting function through the 9 or 15 points centered on the peak ordinate. The cubic interpolation curve is differentiated and the quadratic first derivative function so generated is solved for ν_{max}. A logical section examines the two roots obtained and selects the correct one, if there is ambiguity, or if both roots are imaginary, appropriate alerting messages are printed. Means are provided for printing out details of the various stages of these solutions but they are normally omitted.

The choice between the 9-point and the 15-point cubic convoluting functions in the interpolation step can be made by the operator when the program is initiated, but it cannot be changed from peak to peak during the analysis. The objective is to have the maximum number of points on each side of the peak without extending appreciably beyond the half-bandwidth abscissas. For most spectra of condensed phase systems encoded at 0.5 cm^{-1} intervals the 9-point curve is satisfactory (see note 6).

Similar arguments determine the choice of the 25- or 9-point cubic first derivative convoluting function used to locate where $dT/d\nu$ changes sign. In both these situations the criterion for maximal sensitivity is the ratio of the half-width of the band to the encoding interval.

When the program operates in a mode which locates the absorbance valleys, T is replaced by 1000 - T and the inverted curve is processed; this avoids complications with the sign conventions of the slope within the program. The program output consists of a tabulation of the positions of the interpolated extremae together with their absorbance values and the corresponding transmission as %T. Also designated as "off scale" or "on scale" are the places at which the transmission falls below 3% indicating that an overstrong band is present. When the program is set to locate absorbance minima similar indications are given if the curve dips below 100%T. When both maxima and minima are called for they are presented in two separate tables, the absorbance maxima preceding the absorbance minima. There is no provision for card output.

C. Band Area Integration

The evaluation of the area beneath an isolated absorption band, or set of overlapping bands, involves only the application of an appropriate integration algorithm. In practice the problem is complicated by the confusing lack of uniformity concerning the form and units in which spectroscopists record integrated absorption intensities.

Program XXVI deals with this problem. It uses an integration subroutine (QUADD), taken from a standard software library, to perform the two following basic integrations

$$\text{Area}_1 = \int_{\tilde{\nu}_1}^{\tilde{\nu}_2} \log\left(\frac{P_0}{P}\right)_{\tilde{\nu}} d\tilde{\nu} \qquad (\text{cm}^{-1}) \tag{11}$$

$$\text{Area}_2 = \int_{\tilde{\nu}_1}^{\tilde{\nu}_2} \frac{1}{\tilde{\nu}} \ln\left(\frac{P_0}{P}\right)_{\tilde{\nu}} d\tilde{\nu} \qquad (\text{dimensionless}) \tag{12}$$

It then computes the five integrals (13)-(17) in which the integrated absorption intensities of infrared bands are frequently expressed [24]. The IUPAC Practical Unit

$$\text{Area}_3 = \frac{1}{cd} \int_{\tilde{\nu}_1}^{\tilde{\nu}_2} \log\left(\frac{P_0}{P}\right)_{\tilde{\nu}} d\tilde{\nu}$$

$$= \int_{\tilde{\nu}_1}^{\tilde{\nu}_2} \varepsilon_{\tilde{\nu}}\, d\tilde{\nu} \qquad (\text{mmol}^{-1}\ \text{cm}) \tag{13}$$

The IUPAC Absolute Unit

$$\text{Area}_4 = \frac{1}{Nd} \int_{\nu_1}^{\nu_2} \ln\left(\frac{P_0}{P}\right)_{\nu} \qquad (\text{molecule}^{-1}\ \text{cm}^2\ \text{sec}^{-1}) \tag{14}$$

computes some derived functions of these moments that have value in the analysis of the band shape.

The second, third, and fourth moments of the single-peaked function $f(\nu) = \log(P_0/P)_\nu$ about the ν' ordinate are given by

$$\mu_r = \frac{\int_{-\infty}^{+\infty} (\nu - \nu')^r f(\nu)\, d\nu}{\int_{-\infty}^{+\infty} f(\nu)\, d\nu} \tag{18}$$

where r = 2, 3, 4.

The width of an asymmetric band can be characterized by b_1 and b_2 (Fig. 2) where A and B are points on the curve where $A = 0.5A_{max}$. The half-width $(\Delta\nu_{1/2})$ is $b_1 + b_2$. If the band is symmetric, $b_1 = b_2 = b$ and $\Delta\nu_{1/2} = 2b$. The absolute difference $|b_1 - b_2|$ is a crude measure of the band asymmetry. Where the asymmetry is not excessive we may write $b = 0.5(b_1 + b_2)$ and define a new abscissal scale of "j-units" where

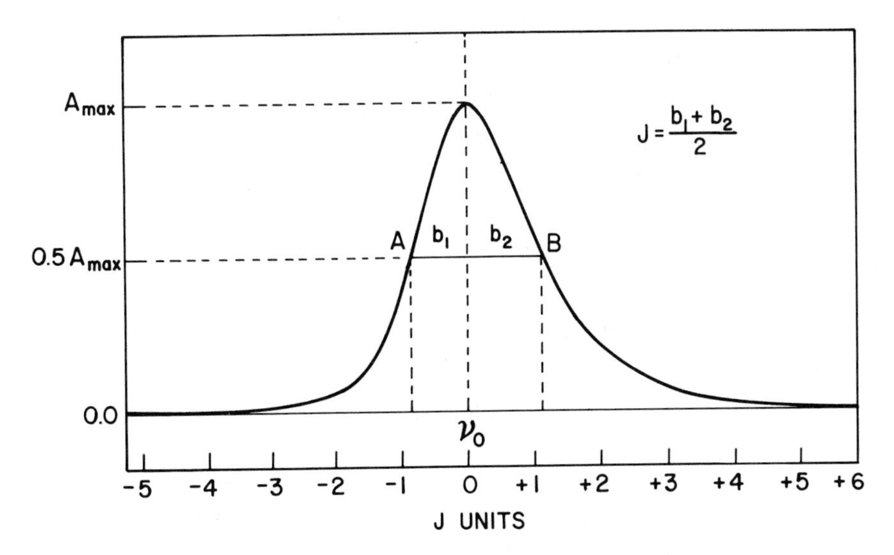

FIG. 2. Program VIII. Characterization of the half-width of an asymmetric band and establishment of the j-unit scale.

$$j = \frac{\nu - \nu_o}{b} \tag{19}$$

in which ν_o is the wave number at the band maximum.

The *truncated* second, third, and fourth moments about ν_o are

$$\mu(j)_r = \frac{1}{b^r} \frac{\int_{-j}^{+j} (\nu - \nu_o)^r f(\nu)\, d\nu}{\int_{-j}^{+j} f(\nu)\, d\nu} \tag{20}$$

These are dimensionless numbers; they are independent of the wave number scale and are therefore pure shape parameters.

Additional quantities of value in the analysis of band assymmetry are $\mu_3(j) \cdot b^3$ and the *kurtosis* $\beta(j)$ where

$$\beta(j) = \frac{\mu_4(j)}{\mu_2(j)^2} \tag{21}$$

Program VIII accepts input as $cm^{-1} \times 10/T \times 1000$ or cm^{-1}/A. All internal operations are performed in cm^{-1}/A units except within the PEAKFIND subroutine. Other necessary input is the starting wave number, the wave number interval, the desired incremental value of j and two data points defining the two ends of the baseline.

The ordinates are first normalized with respect to the linear baseline; this may be chosen to be parallel with or coincident with the abscissal axis, but not necessarily so; the algorithm of Program XXV (cf. Sec. II.E) is used. It is important that both the peak abscissa (ν_o) and the half-bandwidth $(b_1 + b_2)/2$ be determined as accurately as possible. The PEAKFIND algorithm of Program III is called to locate the peak position, which need not necessarily coincide with an encoding wave number. PEAKFIND also checks that the curve is suitable for moment analysis; if more than one peak is located or if both roots of the first derivative of the cubic interpolation curve are imaginary the program is terminated with an appropriate error message; this also occurs if the peak is too intense ($A_{max} > 1.5$).

Having located ν_o, the next step is to evaluate b_1 and b_2. This involves an iterative process; the values of b_1 and b_2 are recorded and also the mean. Neither b_1 nor b_2 need necessarily coincide with an encoding data point; a five-point Lagrange polynomial interpolation function (subroutine TERPOL) is used to locate the $\pm b$ ordinate if this occurs between encoding points. Normally these steps present no difficulties, but the operator has an option of listing the details of the iterative location of b_1 and b_2 if desired. With both b_1 and b_2 established the abscissal scale can now be transposed from cm^{-1} to j-units. This also calls for the use of TERPOL. A table listing the j-value, the corresponding displacement in cm^{-1} from ν_o, and the ordinate intensity in units of A and $T \times 1000$ is next displayed.

The program then enters the final stage of computing the series of integrals to obtain the second, third, and fourth incomplete moments corresponding to each range -j, +j and these, together with the derived parameters $\mu_3 \cdot b^3$ and $\beta(j)$ are tabulated. Card output is also available. The subroutine QUADD, taken from a standard software library, is used for the integration.

F. Fourier Transformation and Evaluation of Time Correlation Curves

The time correlation function $\phi(t)$ is obtained by Fourier transformation of a single-peaked absorbance band. For liquid phase systems it has been interpreted by Gordon [26], Shimazu [27], Bratoẑ, Rios, and Guissani [28], and others in terms of the relaxation kinetics of the damped vibrational and rotational motions of the absorbing molecules due to molecular collision.

$$\phi(t) = \frac{\int_{band} k(\omega)\ \cos[(\omega - \omega_o)t]d\omega}{\int_{band} k(\omega)\ d\omega} \tag{22}$$

where $k(\omega)$ is the absorption index, ω the frequency (rad sec^{-1}) and ω_o the frequency at the band center.

Since

$$k = \frac{2.30258A}{4\pi\tilde{\nu}d} \tag{23}$$

and

$$\omega = 2\pi c\tilde{\nu} \tag{24}$$

where $\tilde{\nu}$ is the wave number (cm^{-1}) and c the velocity of light, Eq. (22) may be written

$$\phi(t) = \frac{\int_{band} (A/\tilde{\nu}) \cos[(2\pi ct(\tilde{\nu} - \tilde{\nu}_o)] \, d\tilde{\nu}}{\int_{band} (A/\tilde{\nu}) \, d\tilde{\nu}} \tag{25}$$

Program XXI evaluates $\phi(t)$ from Eq. (25). The data are accepted as $cm^{-1} \times 10/T \times 1000$ or cm^{-1}/A. The initial steps in the program are similar to those of the preceding Program VIII. The input requires the coordinates of two points forming the ends of the baseline and the ordinates are first normalized with respect to this baseline; commonly, though not necessarily, it will be parallel to or coincident with the abscissal axis. Subroutine PEAKFIND is called to locate the position of the band maximum (ν_o) and to check that the curve is single-peaked, with $A_{max} < 1.5$ and in other ways suitable for the computation of the Fourier transform curve. If found not to be so the computation is terminated with an appropriate error message. The vector of the wave numbers is next adjusted so that ν_o, as found by PEAKFIND, coincides with an encoding data point. If this involves a displacement exceeding 0.05 cm^{-1} the ordinates are also adjusted for the displacement using subroutine TERPOL. If such a displacement has been found necessary, the new wave number and absorbance values are printed.

The program now enters the main section to compute the Fourier transform integrals using subroutine QUADD, and the output is tabulated as t, $\phi(t)$, log $\phi(t)$, δ log $\phi(t)$. The latter is the incremental change in log $\phi(t)$ and is useful in providing a readily recognizable indication of oscillation in the log $\phi(t)$ output. Card output is also available.

IV. INSTRUMENTAL MEASUREMENTS AND CALIBRATION

In this section several programs are described that have been developed to measure some of the instrumental parameters in infrared spectrophotometry. These include the cell path length and the spectral slit width; other programs provide for the automated calibration of the wave number scale.

A. Wave Number Calibration

Programs XXXI-XXXVII concern wave number calibration, but some mention should first be made of the more general aspects of wave number calibration.

In chemical infrared spectrophotometry it is customary to calibrate the wave number scale by recording the spectra of standard materials for which the positions of easily identified band peaks are known with high accuracy. The best sources of such data are the Tables of Wavenumber for the Calibration of Infrared Spectrometers of the International Union of Pure and Applied Chemistry (IUPAC). The first set of IUPAC tables, published in 1961 [29], covered the range 4000 to 600 cm^{-1}; subsequent tables for the far-infrared (600-1 cm^{-1}) were published in 1973 [30] and a second edition of the IUPAC monograph combining both sets of tables with revision of some of the earlier published values is now available [31].

These tables provide a series of spectra adapted for the use of spectrometers of different resolution. Most of the IUPAC tables relate to gas phase spectra. These yield the most accurate calibration values and should always be used where the highest accuracy is demanded; each gas spectrum, however, is suitable only for a narrow range of the spectrum. Most chemical spectroscopists who are concerned with extended wave number range measurements on condensed phase systems will find it more convenient to calibrate with liquid or solid materials since these permit calibration over long ranges of wave number with adequate accuracy; furthermore, there is merit in using a calibrating material which generates a spectrum having bands of the same shape and width as the materials being measured.

Polystyrene film is widely used for this purpose but a more suitable material is a ternary mixture containing 98.6 parts by weight of indene, 0.8 parts by weight of camphor, and 0.8 parts by weight of cyclohexanone (see note 7). This liquid mixture has a very rich spectrum with 77 sharp bands in the mid infrared between 4000 and 600 cm^{-1}. A mixture containing equal parts by weight of the same substances can be used between 600 and 300 cm^{-1}. These spectra are listed in the IUPAC tables and a more detailed discussion has been published [32].

Programs XXXI and XXXII are used to establish and apply a calibration function suitable for use with the conventional double-beam grating spectrophotometers.

1. Generation of Polynomial Wave Number Calibration Functions

The calibration process starts with the measurement of the spectrum of one of the standard calibration materials. This may be either the indene/camphor/cyclohexanone spectrum encoded at 0.5 cm^{-1} intervals or one of the gas phase spectra encoded at 0.1 or 0.2 cm^{-1} intervals.

The digitized output, either as $cm^{-1} \times 10/T \times 1000$ or cm^{-1}/A is processed through Program III or XXX to locate the positions of the absorption maxima (see note 8). Two sets of Hollerith cards are next punched in 8(F.10.2) format; the first lists the true values of the selected calibration points taken from the IUPAC tables, and the second the corresponding values obtained for the measured spectrum by the PEAKFIND algorithm. After checking the input data for compatibility, Program XXXI establishes a vector of the wave number errors ($y = \nu_{obs} - \nu_{cal}$) and uses a least squares library software subroutine to evaluate a set of polynomial error correction functions

$$y_\nu = a_o + a_1\nu + a_2\nu^2 + \cdots + a_n\nu^n \qquad (26)$$

with n ranging from 1 to 5.

Program XXXI also computes the variance of the fit for each power series which it tabulates together with the coefficients $a_o \cdots a_n$. A subroutine ERTEST computes the corrected peak wave numbers (ν_{cor}) and tabulates the true value (ν_{cal}), the experimentally measured value (ν_{obs}), the corrected value (ν_{cor}), the correction ($\nu_{cor} - \nu_{obs}$) and the residual error ($\nu_{cor} - \nu_{cal}$) for each power series. The program also monitors the round-off error in the least squares approximation and gives a warning message if this becomes significant. A set of punched cards listing the polynomial coefficients for each power series is also supplied.

By inspection of this output the operator can decide which of the polynomial functions is most acceptable and the appropriate cards used for the calibration.

2. Application of the Wave Number Calibration

In our laboratory the coefficients of the selected polynomial are fed into our general data clean up program and the output is automatically corrected for wave number errors. Program XXXII performs the same operation.

It accepts the digitized input data from the spectrophotometer as $cm^{-1} \times 10/T \times 1000$ in the standard 8(I5.I4.1X) format together with the appropriate polynomial coefficients. A provision is included for an additional correction due to sinusoidal correlation in the residual error as discussed below, but this will not normally be encountered and is bypassed by an instruction card. Program XXXII computes the wave number correction at each data point. When operating in one mode it will then tabulate the correction, together with the corrected wave number and the corresponding intensity. Punched card output can also be obtained.

For some purposes this wave number corrected output is unsuitable because, if the wave number correction is not constant across the spectrum, the corrected wave numbers will no longer retain the uniform spacing. To circumvent this the operator has the option of running the program in a different mode in which the original wave number values are restored and subroutine TERPOL makes adjustments to the band intensities. These data are tabulated and also punched if required. In a third mode both the wave number adjusted and the intensity adjusted forms of the calibration corrected data are listed.

When operating at wave number intervals of 0.5 cm^{-1}, as would be normal for condensed phase spectra, no difficulties have been encountered due to round-off error. However, when operating with gas phase spectra encoded at 0.1 cm^{-1} intervals, round-off error can cause discontinuities in the wave number vector of the corrected curve. This does not occur if the intensity corrected output is accepted. Appropriate warning messages about this are printed.

3. Correction for Sinusoidal Wave Number Errors

Program XXXIII deals with a situation, fortunately not common with infrared spectrophotometers of commercial manufacture, where, after application of the polynomial correction, the residual error shows systematic variations of an oscillatory character. This is illustrated in Fig. 4 which was obtained during the calibration of a spectrometer over the range $<3400, 3260\ cm^{-1}>$ with gaseous HCN. The spectrum of the calibrating gas is shown in Fig. 3. Curve A of Fig. 4 is the correction ($\nu_{cal} - \nu_{cor}$) and curve B the quadratic calibration curve computed by Program XXXI. It is seen that the distribution of the residual error about the least squares quadratic curve is far from random and is roughly sinusoidal with a periodicity of about 30 cm^{-1}. This corresponds to one turn of the lead screw which drives the grating table and indicates a machining error in the screw pitch. Program XXXIII applies a sinusoidal correction which is superimposed on the polynomial calibration function. It uses the nonlinear least squares band fitting algorithm of Program X (cf. Sec. V.D).

The input to Program XXXIII is the experimentally measured positions of the calibration bands as found by PEAKFIND in Program XXXI (ν_{obs}), the calibration values from the IUPAC tables (ν_{cal}), the coefficients of the selected polynomial correction function, and three parameters defining the sinusoidal error function. These are the estimated amplitude (a), the estimated pitch (p), and one wave number (ν_0) at which the polynomial and sinusoidal curves intersect. The program assumes that the sinusoidal error (δ_s) can be approximated by

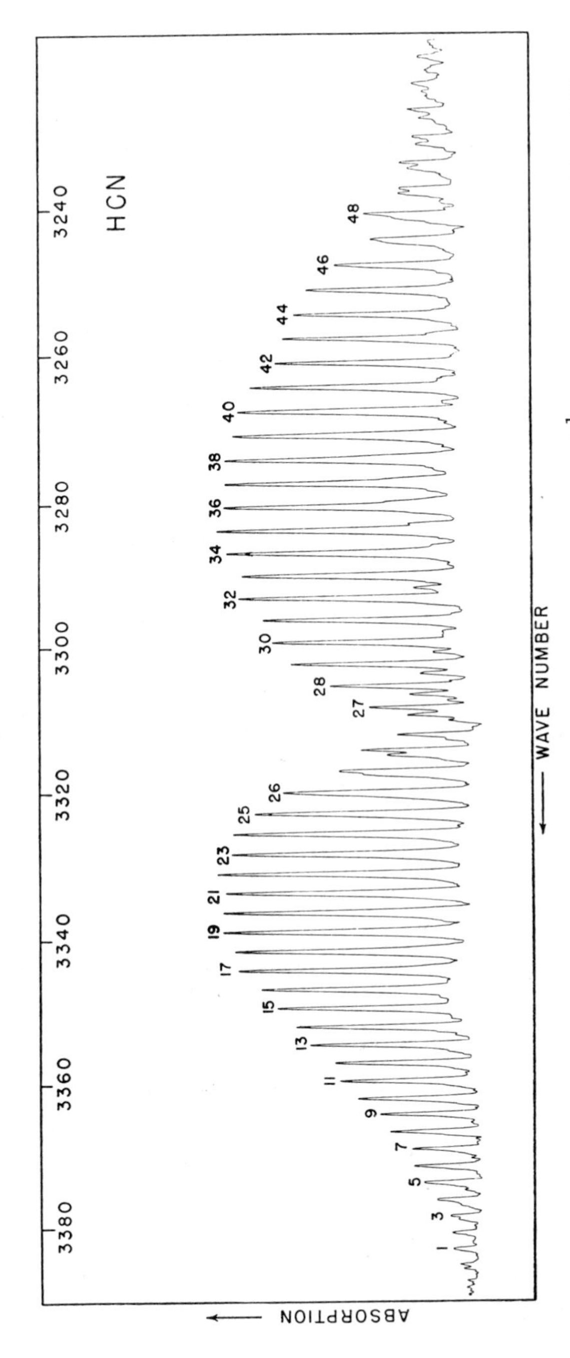

FIG. 3. Hydrogen cyanide calibration curve for the range 3380-3240 cm^{-1} from the IUPAC tables [29].

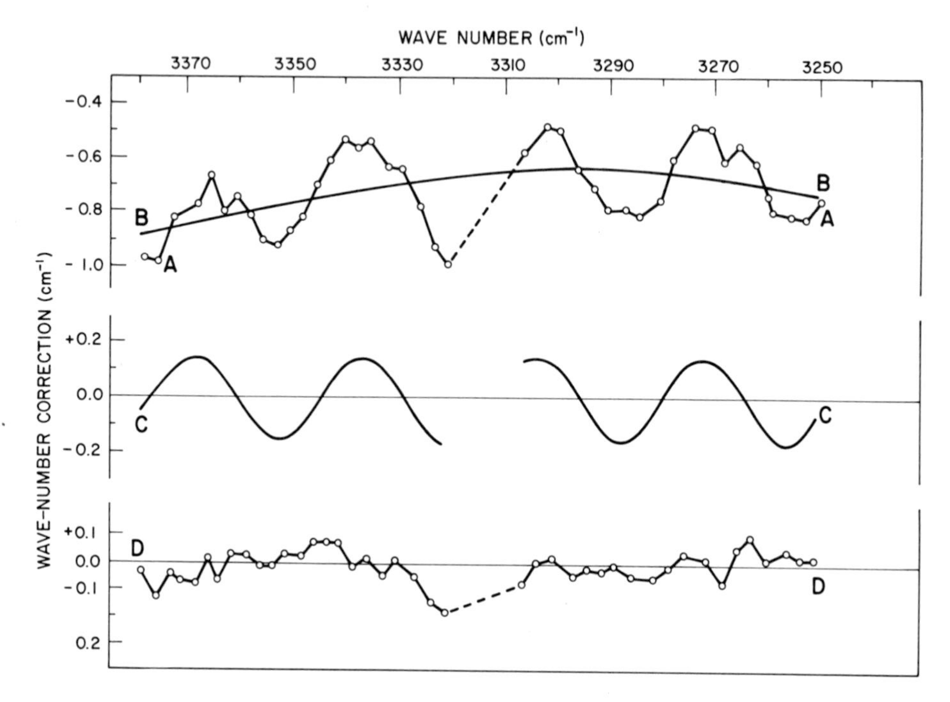

FIG. 4. Program XXXIII. Correction of sinusoidal wave number errors. Curve A: corrections at measured calibration wavelengths. Curve B: quadratic calibration correction curve generated by Program XXXI. Curve C: sinusoidal correction curve generated by Program XXXIII. Curve D: residual differences.

$$\delta_s = a \sin 2 \frac{\pi}{p}(\nu_o - \nu) \tag{27}$$

and optimizes a, ν_o, and p. There is an option to examine the progress of the optimization, but otherwise only the number of iterations and the progress of the convergence are printed. The resultant sinusoidal correction curve is shown in Fig. 4C and the remaining difference after applying this additional correction is shown in Fig. 4D.

In addition to the optimized values of a, ν_o, and p, the printed output from Program XXXIII consists of a tabulation of ν_{cal}, δ_s, the final corrected wave number (ν_{p+s}) and the final residual error ($\nu_{p+s} - \nu_{cal}$). The root mean square of the residuals both before and after the application of the sinusoidal error is also computed.

In the example given here the application of the sinusoidal error correction reduced the rms of the residuals from 0.12 cm^{-1} to 0.06 cm^{-1}. The optimized values of a, ν_o, and p are introduced by a punched card into Program XXXII and they they are present the sinusoidal correction is applied automatically.

4. Calibration of Spectrometers with Constant Angular Motion of the Grating or Prism Table

Programs XXXIV-XXXVII were written to handle data generated on both grating and prism spectrometers using shaft encoders linked through gearing to a rotating shaft turning the grating or prism table with a constant angular velocity. To make such data compatible with these programs it is necessary to establish a calibration curve relating the counter number (m) of the shaft encoder with the wave number and then convert the data encoded at equal counter number intervals (δm) to data encoded at equal wave number intervals ($\delta \nu$).

The input data consist of a vector of calibrating counter numbers (m_1, m_2, $\cdots$, m_n) and a vector of the corresponding wave numbers (ν_1, ν_2, $\cdots$, ν_n). It must be emphasized that the counter number to wave number conversion function may be nonlinear; indeed it will be strongly so for prism spectrometers where it conforms with the dispersion of the prism material which is highly nonlinear at the high wave number end (see note 9).

The input to Program XXXIV consists of the m_{obs} and ν_{cal} vectors which are generated by recording the spectrum of a suitable calibration material and processing the output through Programs III or XXX to locate the peak positions in "m-units." The ν_{cal} vector is obtained from the corresponding calibration peak values taken from the IUPAC tables. Program XXXIV resembles Program XXXI and uses the same library software subroutine to generate a set of least square polynomial calibration functions of the form

$$m_{calc} = a_o + a_1\nu + a_2\nu^2 + \cdots + a_n\nu^n \qquad (28)$$

which, in this series, extends to n = 7.

The program requires that the m's be listed in decreasing numerical sequence but the ν values may increase or decrease. It also locates the m vector nearest to the center and normalizes the m's with respect to this before calling the subroutine to generate the seven sets of polynomial coefficients. Having obtained the coefficients it calculates the computed m's (m_{calc}) for each power series and tabulates m_{obs}, m_{calc}, and the error ($m_{calc} - m_{obs}$) and the standard error of the fit.

The coefficients of the various power series are punched out together with an extra card carrying the normalization factor. The most appropriate power series is selected by the operator and the corresponding coefficients, together with the normalization factor, are used as input to Program XXXV which applies the calibration function to the experimental data.

Program XXXV is similar to Program XXXII but does not incorporate the facility for applying a sinusoidal correction. It accepts the input data in counter numbers and T × 1000 in 8(I5,I4,1X) format. Operating in one mode it converts the counter numbers to wave numbers and tabulates counter numbers, wave numbers, and intensities. In this table the counter numbers will be equally spaced but the wave numbers will not. Operating in the alternative mode the program calls subroutine TERPOL and computes the ordinates corresponding to equal wave number encoding intervals. It tabulates the counter number, the wave number in cm^{-1}, and the intensity in T × 1000. It will also provide card output as $cm^{-1} \times 10/T \times 1000$ in the standard format.

Programs XXXIV and XXXV are used in our laboratory to support a Perkin-Elmer Model 112 spectrometer used for the measurement of attenuated total reflection spectra.

5. Calibration of Prism Spectrometers Using the McKinney and Friedel Algorithm

In automating prism spectrometers for digital output, we have also investigated the use of the Friedel and McKinney function [33] as an alternative to the polynomial power function as a calibration algorithm. This function has also been used by Downie, Magoon, Purcell, and Crawford [34] for the calibration of prism spectrometers and may be written

$$T_\nu = T_o + \frac{B}{\nu_o^2 - \nu^2} + A\nu^2 \tag{29}$$

in which T_ν is the counter reading, T_o a trivial constant determined by the position of the counter zero, A is a correction term which becomes significant only at high wave numbers and is determined by the ultraviolet dispersion characteristics of the prism material and B is a measure of the prism dispersion in the infrared. The ν_o term was originally identified by Friedel and McKinney as a restrahlen factor but Crawford and coworkers treated it as an empirical parameter.

The input to Program XXXVI consists of the arrays of the wave number vector $\nu_1, \nu_2, \cdots, \nu_n$ and the counter number vector $m_1, m_2, \cdots, m_n$ as for Program XXXIV; it uses the algorithm of the band fitting Program X to optimize A, B, and T_o. Initial values of T_o and B are evaluated within the program by dropping the ν^2 term of Eq. (29), which is very small, and solving for T_o and B from two selected values for m and ν chosen from well separated points in the m and ν vector arrays. A nominal value of 10^{-6} is then entered for A and the band fitting subroutine called to optimize A, B, and T_o. The Friedel-McKinney function is then used with the optimized parameters to compute m_{calc} values and the error $(m_{calc} - m_{obs})$ as with Program XXXIV and print the appropriate tabulated output.

In Program XXXVII these Friedel-McKinney coefficients are used to calibrate the experimental data in wave numbers, and, by means of TERPOL yield printed and punched output at equal wave number intervals. No discussion of Program XXXVII is necessary. In practice we have found the Friedel-McKinney algorithm to be less accurate than a fifth, sixth, or seventh order polynomial fit.

B. Computation of Spectral Slit Widths and Slit Programs for Littrow Spectrometers

The resolution of an infrared spectrometer is commonly defined in terms of the half-width of the spectral slit width (s) in cm^{-1}. This is discussed in Ref. 24 where the following expression is derived for a Littrow grating spectrometer

$$s_\nu = \frac{\nu^2 d}{NnF}\left[1 - \left(\frac{n}{2\nu d}\right)^2\right]^{1/2}\left\{w_s + \left[\left(\frac{F}{B\nu}\right)^2 + w_a^2\right]^{1/2}\right\} \tag{30}$$

where ν is the wave number (cm^{-1}), N the number of grating passes, n the order of the spectrum, d the grating spacing (cm), F the focal length of the collimator (cm), w_s the mechanical slit width (cm), and w_a an aberration term.

Program XXVIII tabulates s over a preselected wave number range for a given value of w_s or, operating in a different mode, it tabulates w_s for a given value of s. We have used it more frequently in the second mode to establish mechanical slit schedules to maintain a constant resolution over a chosen range of wave number. This program has no card output.

Program XXIX generates similar tables relating the spectral slit width and the mechanical slit width for a Littrow prism spectrometer on the basis of the equation (see note 10):

$$s_\nu = \frac{\nu}{2N(dn/d\lambda)}\left[\nu\left(1 - n^2 \sin^2 \frac{\alpha}{2}\right)^{1/2} \cdot \frac{w}{2F\,\sin(\alpha/2)} + \frac{1}{p}\right] \tag{31}$$

where N is the number of grating passes, n the refractive index, and $dn/d\lambda$ the dispersion of the prism material, α the apical angle, and p the base length of the prism, and w the mechanical slit width (cm). In Program XXIX the parameters to compute $dn/d\lambda$ for the common prism materials (LiF, CaF_2, NaCl, KBr, CsI) are incorporated in a subroutine. This program has no provision for card output.

C. Correction of the Spectrum for Finite Spectral Slit Distortion

Infrared absorption measurements on condensed phase systems will usually be made under conditions where the resolution is limited by energy considerations. To obtain sufficient energy to provide an adequate detector current the slit function will be determined by the geometry of the slit image [24]. The effect is to diminish the peak intensity and reduce the resolution in inverse proportion to the ratio $s/\Delta\nu_{1/2}$ where $s(\nu)$, the spectral slit function, can be assumed to be triangular, and $\Delta\nu_{1/2}$ is the half-width of the true band.

Mathematically, the effect amounts to a convolution of the true spectrum with the slit function. The continuous slit function can be replaced by a set of equally spaced slit ordinates; for convenience this spacing can be chosen to match the digital encoding interval of the spectrum. To obtain the true band contour from the measured band contour it is therefore necessary to *deconvolve* the measured spectrum with a set of equally spaced ordinates simulating the slit function [21]. A true deconvolution is difficult to achieve; it can be done through Fourier transformation of both the band and the slit function, multiplication of the two transform functions, followed by a reverse Fourier transformation to generate the true absorption curve [24].

The alternative method used in Program VI is a pseudodeconvolution illustrated in Fig. 5. The ordinates of the measured Curve A

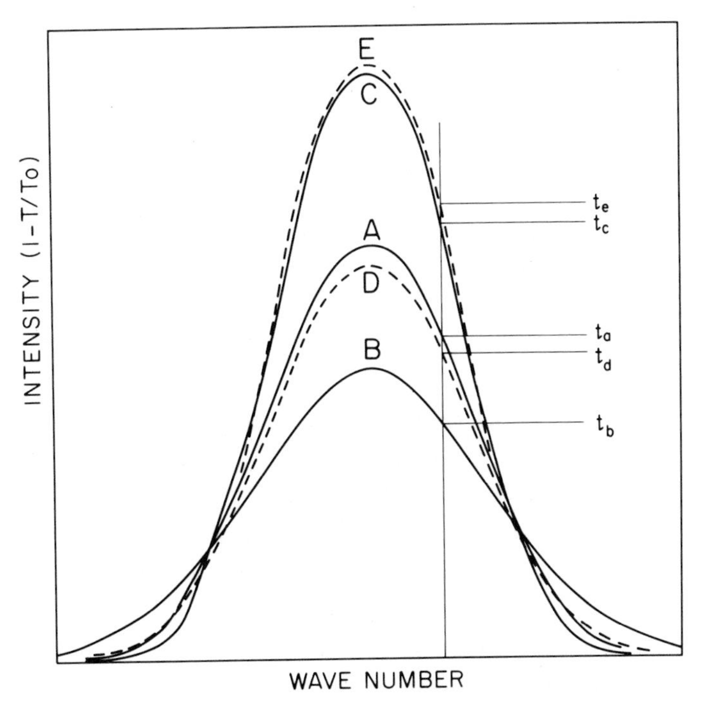

FIG. 5. Program VI. Diagram illustrating the pseudodeconvolution algorithm.

are convolved with the set of slit ordinates to produce curve B. If t_a and t_b are the transmission ordinates at a given wave number ν, curve C is constructed by computing t_c where $t_c = t_a \cdot t_a/t_b$; curve C is the first approximation to the deconvolved band. Curve D is next generated by convolving $t_c(\nu)$ with the slit ordinates. If the initial pseudodeconvolution had been exact, it would be expected that $t_d = t_a$, but this will rarely be so. A second approximation (curve E) is obtained from the ratio $t_e = t_c \cdot t_a/t_d$. This iterative process is continued; normally two to five iterations are required. In each cycle the rms of the ordinate differences (δ) is evaluated where

$$\delta_1 = \frac{1}{N} \sum (t_e - t_c)^2 \qquad (32)$$

and the iteration is continued until δ ceases to decrease, or a preset maximum number of iterations is reached.

It should be observed that δ does not converge; it passes through a minimum value and then increases again. The program monitors δ and stops the computation when δ begins to diverge. This pseudodeconvolution algorithm is based on prior work of Burger and Van Cittert [35] and of A. H. Nielsen, et al. [36] who used a subtractive step $t_c = t_a + (t_a - t_b)$. The reasons for our preferring the product algorithm are discussed in Ref. 7.

Program VI only accepts input data in the $cm^{-1} \times 10/T \times 1000$ units. It operates in two modes. In mode 1 a symmetrical triangular spectral slit function is assumed; for this only the half-width of the slit function need be supplied as the ordinates are computed within the program. Operating in the second mode a set of slit ordinates must be supplied on punched cards. This permits a slit function of any contour, either symmetric or asymmetric to be used.

This program was initially developed in 1965 for use with a small SDS 920 computer with a very limited storage capacity. It was therefore necessary to process the data in sections of 160 data points. Under these circumstances the loss of several data points at each end of the processed data sections on each convolution was serious and a method was evolved to recover these by means of an adjustment matrix. This step is retained in the present program though it is of less relevance now that the program will normally be operated on computers having sufficient storage capacity to deal with the complete spectrum in a single computation.

With spectra having a high noise level, difficulties have been encountered due to Slutzky-Yule oscillation (see note 11). In general the deconvolution tends to increase the apparent noise levels progressively with each iteration cycle. To control this oscillation the input data are first smoothed with a seven-point quartic polynomial smoothing function and a similar smoothing of the data occurs before entering each iteration cycle. Card output in the standard $cm^{-1} \times 10/T \times 1000$ units is provided if required. The convolution algorithm is discussed further in connection with Program V in Sec. V.C.4.

D. Evaluation of Cell Path Length

The measurement of the path length of the absorption cell is critically important in accurate spectrophotometry. Three methods are available: (i) a traveling micrometer can be used to focus successively on the two inner faces of the cell; (ii) absorption measurements can be made on a standard material of known absorbance; (iii) interferemetric methods can be used, based on the analysis of the fringes observed when the empty cell is scanned over a suitable wave number range. Method (iii) is the most accurate provided the cells are of good optical quality.

Programs XXII and XLV both analyze these interference fringes. Program XXII is suitable for cells having path lengths exceeding 100 um and Program XLV has been developed more recently for thin cells with path lengths in the range 1-100 μm.

1. Path Lengths of Cells of Medium Thickness (100-1000 μm)

For Program XXII an interference pattern covering about 20 successive fringes is required with the data encoded at intervals in the range 0.2-0.5 cm^{-1} to record about 500 points. For a cell of 1 mm path length this will incur a scan of 110 cm^{-1}. The program accepts the data in the standard $cm^{-1} \times 10/T \times 1000$ format. The PEAKFIND algorithm of Program III is first called to locate the maxima and minima which are stored in separate arrays. The central (kth) minimum is located where, if the total number of bands be j, k = j/2 if j is even and 1 + (j - 1)/2 if j is odd.

A first value for the path length is computed from the equation

$$d_1 = \frac{(k - 1)}{2n(\nu_k - \nu_1)} \tag{33}$$

where d_1 is the path length (cm) identified with ν_1, and n is the refractive index of the material in the cell (a value of 1.00026 for air at 25° is normally assumed). A second value for d is next calculated by replacing ν_1 by ν_2 and ν_k by ν_{k+1} for the range from the second to the k + 1th fringe. This process is continued through the fringe pattern to obtain a set of k values for d if j is odd, or k + 1 if j is even. The same procedure is repeated for the fringe minima; separate averages are obtained for the values of d derived from the maxima and from the minima and an overall average $\bar{d}$ by further averaging of these two. The fringe order N_m is also computed for each extremum from the equation

$$N_m = 2\bar{d}\nu_m n \tag{34}$$

The program tabulates the wave number positions and intensities of the maxima and minima with the corresponding values of N_m and, in a separate table, it lists the values obtained for the path length and their averages.

For cells giving well defined fringe patterns, N_m will normally diminish by unit intervals as the wave number decrease, though the values may not be integral as simple optical theory would predict. The reasons for this are considered in the discussion of Program XLV below.

2. Path Lengths of Thin Cells (1-100 μm)

Recently our laboratory has become involved with the measurement of the infrared spectra of pure liquids for which cells with path lengths in the range 1-100 μm are needed [14,37-39]. As the path length diminishes it becomes increasingly difficult to maintain the windows parallel. The cells are assembled under sodium light and the pressures on the clamping screws adjusted so that the internal fringe pattern in the cavity which is seen by reflection consists of parallel bars perpendicular to the direction of the spectrometer slit. This assures that the cell cavity has a trapezoidal cross section with a mean thickness of d_o at the center of the working area and $d_o - \Delta$ and $d_o + \Delta$ at the optically significant window extremities as illustrated in Fig. 6. When such cells are used, the data reduction to obtain the true absorption takes account of the variation in path length across the cell (cf. Program XLVI; see Sec. VI.C).

When such an empty wedge shaped cell is scanned on the spectrophotometer the fringe pattern observed will no longer be sinusoidal. In Program XLV a band-fitting subroutine is used to optimize the fit of the fringe pattern and evaluate both d_o and Δ. The program also takes account of the convergence of the incident beam, the radiation incident on the cell extending over a range of incidence from zero to $\theta°$ (Fig. 6). It also has provision to take account of the polarization discrimination of the spectrophotometer though that is normally not important.

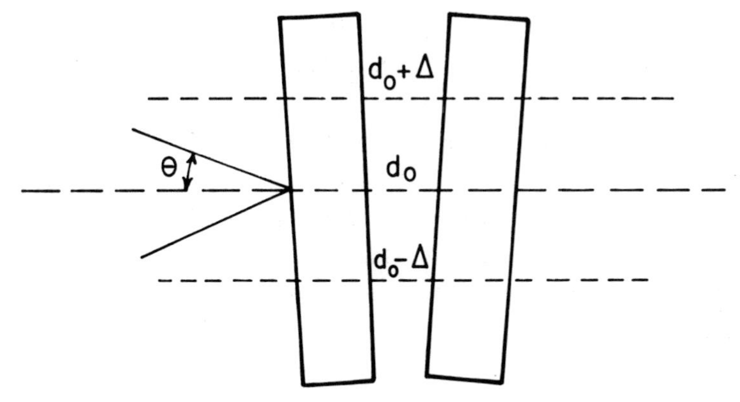

FIG. 6. Nonparallel cell assembly in a convergent beam (diagrammatic).

Program XLV operates in three modes. In mode 1 it evaluates d_o and Δ, assuming $\theta°$ to be known. In mode 2 it evaluates d_o, Δ, and $\theta°$. In mode 3 it evaluates d_o, Δ, and a surface roughness factor (σ) to take account of the imperfect polish of the cell windows. This latter factor is not significant when the cell is used in the mid and far-infrared region ($\nu < 4000\ cm^{-1}$). However in the fringe pattern analysis of cells thinner than 2 μm it is necessary to measure the fringes in the visible and near infrared since their spacing is too great to provide the necessary number of data points in the mid-infrared. In the visible and near-infrared the correction for the imperfect reflectivity of the cell surfaces is significant.

The input to Program XLV consists of the digitized fringe pattern, the refractive index dispersion data for the window material and the polarization coefficients of the spectrophotometer if polarization discrimination is to be considered. Estimated input values are entered for d_o, Δ, $\theta°$, and, if the measurements extend into the near infrared, for the surface roughness parameter σ. The program calculates the reflection losses of the radiation in transmission through the cell (cf. Program XLIV, Sec. VI.B) and compares them with observed fringe data. It uses a nonlinear least squares procedure to adjust the values of d_o, Δ, $\theta°$, and σ (or such of them as are called for by the program mode), and optimizes the fit to the experimentally observed fringe pattern. In this program the optimization algorithm developed by M. J. D. Powell [41] is used. This approximates the partial derivatives by a numerical process whereas those used in our optimization of infrared absorption bands require that the partial derivatives be expressed explicitly in terms of the equation of the fitted curve. The output of the program consists of (i) a tabulation of the process of the optimization, (ii) the final values obtained for the requested optimized parameters, and (iii) a tabulation of the true and computed ordinates of the fringe pattern.

E. Calculation of Infrared Dispersion Data for Optical Materials

In studies of the infrared transmission of thin liquid films and in quantitative infrared reflection spectrophotometry, our laboratory has needed quickly accessible data on the refractive index of a variety of optical materials over extended ranges of the infrared spectrum. In Program XXXVIII we have assembled dispersion data on 21 materials that are used for infrared cell windows [41-54]. This program will generate tables of the refractive indices of these materials over preselected wave number ranges. The materials and wave number ranges are listed in Table 3.

In the literature three forms of empirical dispersion equations occur

$$n_\lambda = A + \frac{B}{\lambda^2 - 0.028} + \frac{C}{(\lambda^2 - 0.028)^2} + D(\lambda^2 - 0.028) + E(\lambda^2 - 0.028)^2 \qquad (35)$$

$$n_\lambda = \left[1 + \lambda^2\left(\frac{A}{(\lambda^2 - B)} + \frac{C}{(\lambda^2 - D)} + \frac{E}{(\lambda^2 - F)} + \frac{G}{(\lambda^2 - H)} + \frac{P}{(\lambda^2 - Q)}\right)\right]^{1/2} \qquad (36)$$

$$n_\lambda = \left[A - B\lambda^2 - C\lambda^4 + \frac{D}{(\lambda^2 - E)} + \frac{F}{(\lambda^2 - G)} + \frac{H}{(\lambda^2 - P)}\right]^{1/2} \qquad (37)$$

where n is the refractive index at the wavelength λ (μm) and A-H, P, Q are numerical constants.

Program XXXVIII requires as input the wave number range and wave number interval and also a code number identifying the material. It selects the appropriate interpolation function and the constants which are stored in a subroutine DATA. It checks that the requested wave number range is consistent with the optical properties of the material, computes the refractive indices and tabulates the output. The indices are computed at preselected wave number intervals (which must not be less than 0.5 cm^{-1}) to 5 decimal places, but they are truncated to 4 decimal places in the

TABLE 3

Dispersion Data on Infrared Optical Materials in Program XXXVIII

Material	Temp. (°C)	Wave number range (cm^{-1})	Error (× 10^4)	Ref.
Air	25	10000 - 10	1	42
Germanium	25	5000 - 741	6	43
Irtran-2 (zinc sulfide)	24	10000 - 741	3	43
Calcium aluminate	24	16667 - 2326	3	43
Lithium fluoride	24	20000 - 1667	3	43
Irtran-1 (magnesium fluoride)	24	10000 - 1493	3	43
Magnesium oxide	24	20000 - 1818	3	43
Silicon	24	7693 - 909	3	43
Fused quartz	24	20000 - 2326	3	43
Strontium titanate	24	10000 - 1887	3	43
KRS-5 (thallium bromide iodide)	24	17331 - 254	4	44
Sapphire (aluminum oxide)	24	37708 - 1793	1	45
Arsenic trisulfide	24	16667 - 834	2	46
Barium fluoride	25	37708 - 967	1	47
Calcium fluoride	24	43706 - 1029	1	48
Cesium iodide	24	33701 - 189	4	49
Silver chloride	23.9	20000 - 488	5	50
Cesium bromide	27	27396 - 255	4	51
Potassium bromide	22	24712 - 398	2	52
Potassium chloride	18	54054 - 449	2	53,54
Sodium chloride	18	54054 - 600	2	53,54

tabulated output with round-off adjustment. They are listed only at the nearest requested wave number at which the fourth decimal place changes. This condenses the table without loss of significant information; the fifth decimal place, if required, can be determined by the operator by linear interpolation. It should be noted that the refractive indices of some of these materials may vary by as much as 0.05 between different melts [43]. The data are computed for the temperature listed in Table 3. The variation of the refractive index with temperature is generally less than 0.001 per °C but there are exceptions, notably KRS-5 for which the temperature gradient is ~0.01. This program has no provision for card output.

V. BAND SIMULATION AND BAND CONTOUR ANALYSIS

This section deals with a series of programs to simulate infrared absorption bands and band envelopes and to optimize their fit to experimental curves.

A. Mathematical Representation of Infrared Bands

Several mathematical functions are used to represent the shapes of infrared absorption bands of condensed phase systems. The most common is the Cauchy function which is the basis of the Lorentz curve. For a single isolated band this is usually written

$$A(\nu)_c = \frac{a}{b_c^2 + (\nu - \nu_o)^2} \tag{38}$$

where A_ν is the absorbance at wave number ν, ν_o the wave number of the peak absorbance, 2b the width of the band at half-maximal intensity, and a/b^2 the peak absorbance.

The other widely used shape function is the Gauss curve

$$A(\nu)_g = \frac{a}{b_g^2} \exp\left[-(\nu - \nu_o)^2 \frac{\ln 2}{b_g^2}\right] \tag{39}$$

The physical factors determining the true band shape and its relation to these functions have been reviewed [24].

For convenience in dealing with band fitting algorithms, it is preferable to write Eqs. (38) and (39) as

$$A(\nu)_c = x_1\left[1 + x_3^2(\nu - x_2)^2\right]^{-1} \tag{40}$$

$$A(\nu)_g = x_5 \exp\left[-x_4^2(\nu - x_2)^2\right] \tag{41}$$

where x_1 is the peak absorbance of the Cauchy curve and x_5 the peak absorbance of the Gauss curve, x_2 the wave number at the peak, $x_3 = b_c^{-1}$ and $x_4 = \ln 2(b_g^{-2})$.

It is found empirically that a better approximation to the true shapes of isolated infrared bands of condensed phase systems is given by some function having the predominant character of $A(\nu)_c$ perturbed by a small $A(\nu)_g$ element [55,56]. This can be achieved with a Cauchy-Gauss product function

$$A(\nu)_p = x_1\left[1 + x_3^2(\nu - x_2)^2\right]^{-1} \exp\left[-x_4^2(\nu - x_2)^2\right] \tag{42}$$

or a Gauss-Cauchy sum function (see note 12):

$$A(\nu)_s = x_1\left[1 + x_3^2(\nu - x_2)^2\right]^{-1} + x_5 \exp\left[-x_4^2(\nu - x_2)^2\right] \tag{43}$$

In $A(\nu)_p$ x_1 is the peak height of the composite function but in $A(\nu)_s$ x_1 is the peak height of the Cauchy component and x_5 the peak height of the Gauss component. Comparing Eqs. (40)-(43) it is seen that the optimization of the fit of $A(\nu)_c$ of $A(\nu)_g$ requires the adjustment of three parameters (x_1, x_2, x_3, or x_5, x_2, x_4). For $A(\nu)_p$ four parameters are involved (x_1, x_2, x_3, x_4) while for $A(\nu)_s$ there are five (x_1, $\cdots$, x_5). With the increase in the number of adjustable parameters the potential for a more accurate fit to the experimental band contour increases, but so also does the amount of computation involved and, more significantly, the likelihood of their being multiple solutions within the tolerated limits of accuracy.

A modified Cauchy-Gauss sum function which reduces the number of adjustable parameters to four can be obtained by fixing the ratio of the half-widths of the Cauchy and Gauss components. This is done in the *restricted Cauchy-Gauss sum function*

$$A(\nu)_{s*} = x_1\left[1 + x_3^2(\nu - x_2)^2\right]^{-1} + x_5 \exp\left[-kx_3(\nu - x_2)^2\right] \tag{44}$$

where $x_4 = kx_3$ in which k is a prefixed quantity. Comparative studies on simulated spectra [56] indicate that in many cases $A(\nu)_{s*}$ with k in the range $\langle 0.7, 0.8 \rangle$ will yield a superior fit to the band contour than $A(\nu)_p$. There are however intrinsic difficulties with the sum function due to the fact that mathematically acceptable fits can be obtained with either x_1 or x_5 negative, indicating that the fit is obtained by subtracting one component from the other. Such a solution can obviously have little physical relevance. The optimization programs for $A(\nu)_s$ and $A(\nu)_{s*}$ discussed later in this section make provision for the operator to hold the signs of x_1 and x_5 positive, but in some cases this fails and the program converges to a solution having x_1 or x_5 near zero and a solution is obtained based on the simpler $A(\nu)_g$ or $A(\nu)_c$ functions. In all the optimization programs an additional degree of freedom is present since the baseline is allowed to float parallel to the abscissal axis.

B. The Damped Harmonic Oscillator Function

In more theoretically based studies of infrared transmission and reflection spectra the damped harmonic oscillator function has advantages over the Cauchy-Gauss functions for the description of band envelopes of condensed phase systems [38,58-60]. It is appropriate to discuss it here as Program XLII generates single bands and overlapping band systems based in this function. We have not yet developed optimizing band fit programs which use it.

If the complex refractive index of the absorbing material is written

$$\hat{n} = n + ik \tag{45}$$

where n is the refractive index and k the adsorption index (see note 13), it follows from electromagnetic theory that if the electric vector of the radiation be written $E = E_0 \exp -2\pi ic\nu t$, where ν is the wave number and c the velocity of light

$$k_\nu = \frac{\ell n(P_0/P)_\nu}{4\pi\nu d} = \frac{2.30258A_\nu}{4\pi\nu d} \tag{46}$$

In evaluating the optical behavior of thin layers of absorbing materials, and in studies of attenuated total reflection, it is more convenient to express band intensities in terms of k_ν than A_ν. Furthermore, if the complex dielectric constant of the absorbing material be written

$$\hat{\varepsilon} = \varepsilon_1 + i\varepsilon_2 \tag{47}$$

and it be assumed that the molecule, in its interaction with the radiation, behaved like a damped harmonic oscillator, then for an isolated absorption band

$$\hat{\varepsilon}_\nu = \varepsilon_\infty + \frac{S^2}{(\nu_o^2 - \nu^2) + i\nu\gamma} \tag{48}$$

where S^2 is the oscillator strength, ν_o the resonant wave number, and γ a damping constant. (S^2, ν_o, and γ all have dimensions of cm^{-1} and k is dimensionless.)

From dispersion theory the following relations are derived between ε_1, ε_2, n, k, S^2, ν_o, and γ for isolated infrared absorption bands. These are important in the algorithm of Programs XLII-XLVI and in studies of the transmittance and reflectance behavior of infrared radiation, especially where Kramers-Kronig transformations are involved [38,39].

$$\varepsilon_{i(\nu)} = n^2 - k^2$$

$$= n_\infty^2 + \frac{S^2(\nu_o^2 - \nu^2)}{(\nu_o^2 - \nu^2)^2 + \nu^2\gamma^2} \tag{49}$$

$$\varepsilon_{2(\nu)} = 2n\nu k = \frac{S^2\gamma}{(\nu_o^2 - \nu^2)^2 + \nu^2\gamma^2} \tag{50}$$

$$n_\nu = \left[\frac{1}{2}\left(\sqrt{\varepsilon_{1(\nu)}^2 + \varepsilon_{2(\nu)}^2} + \varepsilon_{1(\nu)}\right)\right]^{1/2} \tag{51}$$

$$k_\nu = \left[\frac{1}{2}\left(\sqrt{\varepsilon_{1(\nu)}^2 + \varepsilon_{2(\nu)}^2} - \varepsilon_{1(\nu)}\right)\right]^{1/2} \tag{52}$$

C. Programs to Generate Simulated Bands and Band Envelopes

Programs XIII, XIV, XVI, XIX, XX, and XLII generate and analyze simulated isolated bands or band envelopes based on the functions discussed above. Program VII is a general purpose program which will generate a single band envelope of any explicit single peaked function $A_\nu = f(\nu)$; Program V permits the controlled introduction of asymmetry into otherwise symmetrical bands.

1. The Cauchy-Gauss Functions

The input to Program XIV is a vector of parameters (x_1, x_2, x_3, x_4), (x_2, x_3, x_4, x_5), or (x_1, x_2, x_3, x_5) of Eqs. (40)-(42) and (44) together with the constant baseline displacement (α). It will generate the functions $A(\nu)_c$, $A(\nu)_g$, $A(\nu)_p$, or $A(\nu)_{s*}$. The wave number range and encoding interval must be specified and, for $A(\nu)_{s*}$ the proportionality constant k. The output may be obtained as absorbance or T × 1000. For an isolated band the input parameters are entered four on a single card; for overlapping band envelopes the parameters of each band are entered, one set per card, to a maximum of 20 bands. Punched card output is available in the standard formats for cm^{-1} × 10/T × 1000 or cm^{-1}/A. When operating in the $A(\nu)_p$ mode, pure Cauchy functions are generated by entering $x_4 = 0.0$ and pure Gauss functions by entering $x_3 = 0.0$. Program XX will deal in a similar manner with $A(\nu)_s$ with input vectors $x_1, \cdots, x_5$, one set per card to a maximum of 20 cards, together with α.

The half-bandwidths of these functions are computed by Programs XIII and XIX, where XIX is a variant of XIII written specifically for $A(\nu)_s$. These programs also evaluated the empirically defined Cauchy-Gauss proportionality ratios S_p and S_s.

For $A(\nu)_c$ and $A(\nu)_g$ the half-bandwidths $\Delta\nu_{1/2(c)}$ and $\Delta\nu_{1/2(g)}$ are given by simple explicit functions

$$\Delta\nu_{1/2(c)} = \frac{2.0}{x_3} \tag{53}$$

$$\Delta\nu_{1/2(g)} = \frac{2\sqrt{\ln 2}}{x_4} \tag{54}$$

The same holds for the band areas

$$\int_{\text{band}} A(\nu)_c \cdot d\nu = \frac{\pi}{2} \cdot x_1 \cdot \Delta\nu_{1/2(c)} = \frac{\pi x_1}{x_3} \tag{55}$$

$$\int_{\text{band}} A(\nu)_g \cdot d\nu = \frac{\sqrt{\pi}}{2\sqrt{\ln 2}} \cdot x_5 \cdot \Delta\nu_{1/2(g)} = \frac{\sqrt{\pi}\, x_5}{x_4} \tag{56}$$

There are no corresponding simple expressions for $\Delta\nu_{1/2(p)}$, $\Delta\nu_{1/2(s*)}$ or $\Delta\nu_{1/2(s)}$. These quantities must be computed by iterative methods that are too complicated to discuss here [8,9].

The Cauchy-Gauss ratio of $A(\nu)_p$ is defined empirically as

$$S_p = \frac{|x_3|}{|x_3| + |x_4|} \tag{57}$$

which takes a values of 1.0 for a pure Cauchy and 0.0 for a pure Gauss curve: the absolute values x_3 and x_4 are used because in the optimization band fit algorithms x_3 and x_4 occasionally appear with a negative sign (see note 14). S_s and S_{s*}, the Cauchy-Gauss ratios for the sum function, are defined in terms of the relative areas beneath the Cauchy and Gauss component bands

$$S_s = \frac{\int_{\text{band}} A(\nu)_c \, d\nu}{\int_{\text{band}} A(\nu)_c \, d\nu + \int_{\text{band}} A(\nu)_g \, d\nu} \tag{58}$$

which also takes the value 1.0 for the pure Cauchy and 0.0 for the pure Gauss curve.

Program XIII accepts as input the appropriate sets of x vectors and tabulates them together with S_p or S_s. For $A(\nu)_p$ it lists the half-widths of the composite band, while for $A(\nu)_s$ it also lists the half-widths of the Cauchy and Gauss components. In the iteration algorithms to obtain $\Delta\nu_{1/2(p)}$, $\Delta\nu_{1/2(s*)}$, and $\Delta\nu_{1/2(s)}$ a quadratic equation must be solved yielding two possible solutions. Normally one root is imaginary but the program will list both values if they are real.

The introductory description to Program XIII contains a table relating the half-bandwidths, shape ratios, and the corresponding values of x_3, x_4, and k over a wide range. This is useful in setting up optimization programs and making quick assessments of the half-width and shape ratios without resort to the program.

2. The Damped Harmonic Oscillator Program

Program XLII generates a single damped harmonic oscillator band or overlapping band envelopes up to a maximum of 10 components. For each component the input data are a single card carrying a vector of the three parameters S^2, ν_o, and γ of Eqs. (48)-(52) with the refractive index (n_∞) on a separate card. The beginning and ending wave numbers of the range and the encoding interval are also entered. In the program $\varepsilon_{1(\nu)}$ and $\varepsilon_{2(\nu)}$ are first evaluated from Eqs. (49) and (50) and from them ν and k_ν by Eqs. (51) and (52). The printed output tabulates ν, $\varepsilon_{1(\nu)}$, $\varepsilon_{2(\nu)}$, n_ν, and k_ν. Card output of ν, $\varepsilon_{1(\nu)}$, $\varepsilon_{2(\nu)}$ or ν, k_ν, and n_ν is available, or both sets of quantities are available in two separate card decks.

3. The Generalized Function $A_\nu = f(\nu)$

Program VII generates a set of band ordinates for any single valued explicit function $A_\nu = f(\nu)$. It does not deal with overlapping band systems. The required input is the starting wave number, the wave number interval, and the number of ordinates. The function is entered on a separate card, and must be expressed in terms of absorbance. The printed output is a tabulation of either cm^{-1}/A or $cm^{-1} \times 10/T \times 1000$; the corresponding card outputs are available in the standard formats.

4. Asymmetric Bands

The bands generated by any of the Cauchy-Gauss functions discussed above are strictly symmetric with respect to the ν_o ordinate (see note 15). However, controlled asymmetry can be induced by convolving these ordinates with an asymmetric function. Program V provides a convenient means of doing this.

Program V was written in the course of developing the pseudodeconvolution algorithm of Program VI [22]. It convolves a set of transmission ordinates with a set of (slit) ordinates, a restriction being that the (slit) ordinates must be equally spaced with the same wave number interval as the transmission ordinates. The program operates in two modes; in the first a symmetrical triangular convolving function is assumed, the half-width of the triangle is

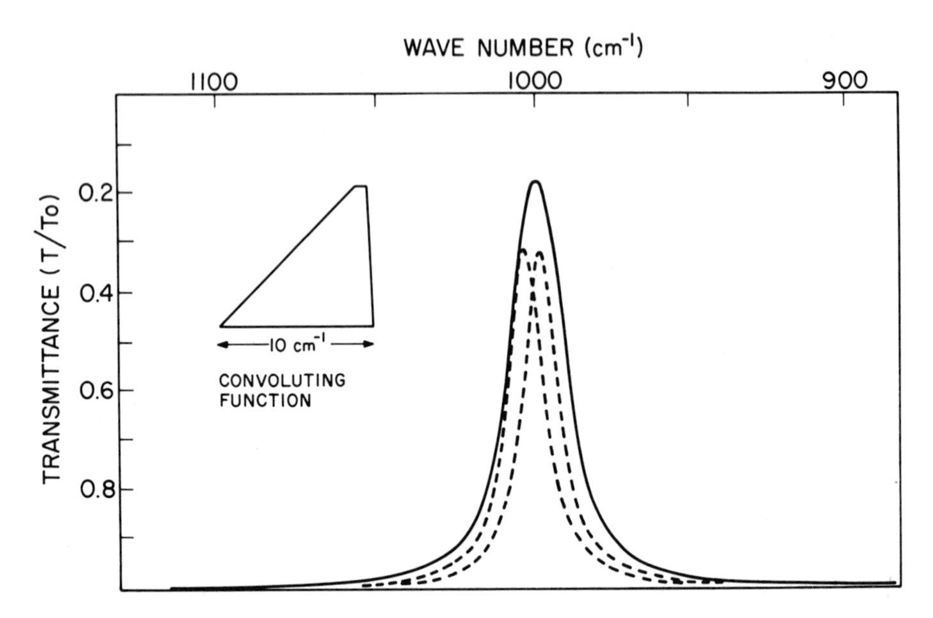

FIG. 7. Program V. Illustration of the use of the slit function convolution algorithm to generate band envelopes with controlled asymmetry.

supplied and the ordinates of the convolving function are computed within the program. In the second mode, which is relevant to the application considered here, any set of suitably spaced convolving ordinates can be introduced on punched cards to a maximum of thirty-five. Input and output to this program are restricted to units of $cm^{-1} \times 10/T \times 1000$ but modification to deal with absorbance ordinates would involve only trivial changes. Card output in the standard format is available.

An illustration (see note 16) of the application of Program V to the generation of an asymmetric band is shown in Fig. 7 where a slightly asymmetric single peaked band was obtained by the addition (in absorbance) of a symmetrical pair of component bands followed by convolution with a highly asymmetrical trapezoidal function [55].

D. Optimization of Band Fits

Program X is the basic program utilizing the algorithm developed in our laboratory by J. Pitha for fitting Cauchy-Gauss functions to experimental band envelopes. Programs XI, XII, XVI, XVII, and XVIII are variants that can be dealt with more briefly.

The mathematical techniques of nonlinear least squares approximation have been widely studied since the advent of the electronic computer; much of the earlier work was stimulated by the importance of these methods in automated lens design. The first applications of nonlinear least squares techniques to the analysis of spectroscopic band shapes were made in 1962 by Stone [61] and shortly afterward by Papoušek and Plivá [62] in the infrared and by Bell and Biggers in the visible and ultraviolet [63]. In our laboratory we began to investigate several methods in 1966 with the initial objective of compressing the storage of spectral data [64], but subsequently the techniques have been used more extensively in analyzing overlapping band systems. In our initial development of the algorithm incorporated in Program X we examined seven nonlinear least squares procedures described in the literature [61,65-69] and compared their efficiency when applied to infrared band contours of increasing complexity from a single simulated band to a 16-band section of the spectrum of a steroid. The technical differences and a comparison of their efficiency, as judged both by the accuracy and rate of convergence, are discussed in Ref. 55. The algorithm finally incorporated into Program X has since been used widely and we have found no need to make modifications. The main feature contributing to its efficiency is the method of selecting the optimal value of the damping constant p which is described in an appendix in Ref. 55.

Program X will fit infrared band contours to envelopes containing a maximum of 20 component bands. In one mode it operates with $A(\nu)_c$, $A(\nu)_g$, and $A(\nu)_p$ and in the other mode with $A(\nu)_{s*}$. The input is the vector of experimental ordinates in units of T × 1000 and the corresponding equally spaced wave numbers as $cm^{-1} \times 10$.

This is followed by the set of 4m band parameters where m is the number of bands; these will be x_1, x_2, x_3, 0.0 for $A(\nu)_c$, x_1, x_2, 0.0, x_4 for $A(\nu)_g$, x_1, x_2, x_4 for $A(\nu)_p$, and x_1, x_2, x_3, x_5 for $A(\nu)_{s*}$. An additional card carries the input value of the baseline displacement (α) and, for $A(\nu)_{s*}$, another card lists the preset constant x_3/x_4 ratio k. Additional input information is the set of three preset constants UP, DOWN, and CHANGE which control the loop optimizing the damping factor p. Other necessary inputs are the operating instructions which allow the options to restrict the optimized bands to widths in the range 1-50 cm^{-1} and to hold x_1 and x_5 to positive or zero values. The operator can also control the amount of information reported about the progress of the iterations. This will normally be limited to a few warning statements if difficulties occur, but if required extensive information can be extracted about the solution of the various matrix equations that are involved.

After each iteration the computed transmission ordinates are compared with the experimental transmission ordinates and the minimization function (FSM) evaluated

$$FSM = \sum^{n} f_i^2 \tag{59}$$

where

$$f_{i(\nu)} = \left(\frac{P}{P_0}\right)_{\nu(obs)} - \left(\frac{P}{P_0}\right)_{\nu(calc)} \tag{60}$$

and n is the number of ordinates.

The program also records FM and WFM, the maximal value of f_i and its wave number, together with other diagnostic information about the progress of the convergence. The computation terminates when one of the following conditions arises: (i) FSM falls below the lowest experimentally significant value (this is preset in the program at 0.001 corresponding to 0.1%T); (ii) the fractional change in FSM from the preceding cycle is less than 1.002, which indicates that the convergence limit has been reached; (iii) the preset number of iteration cycles has been completed; (iv) a preset maximum computation time has elapsed. On termination the optimized values of the 4m x parameters are tabulated together with the optimized value of the baseline displacement α and the relevant statistical information discussed above. On request the optimized x parameters and α will be punched, 4 per card in a format compatible with input to Programs XIII, XIV, and XV.

The use of the unrestricted sum function, $A(\nu)_s$, is provided for in Program XVI but is dimensioned only for a maximum of 10 bands.

The maximum number of component bands that can be dealt with by Program X is 20. This is set by the size of the matrices which must be inverted many times during the course of the computation. In Program XI a "multiplet version" of the same algorithm is used. In principle this can deal with spectra of unlimited length and complexity. In operation the program accepts the complete vector of the experimental ordinates and the complete set of the guessed parameters. In the optimization section only a maximum of five bands is dealt with in one cycle and there are two iterations. The program then drops all but the lowest wave number band from the operational subspace and picks up the next consecutive set, retaining one overlap band, and optimizes the fit of this section. This process continues until the complete spectrum has been traversed when the operational subspace moves back to the high wave number end and the progression through the spectrum is repeated. The same limiting controls apply to terminate the program as for Program X. In Program XI only $A(\nu)_c$, $A(\nu)_g$, and $A(\nu)_p$ functions can be used, but Program XVIII deals in a similar manner with $A(\nu)_s$. The comparative efficiency of Programs X and XI is discussed in Ref. 55 where it is surmised that the advantage of XI over X will occur where the number of bands to be optimized exceeds about 17.

Program XVII is a simpler version of the basic Program X designed to operate only with the $A(\nu)_c$ function. It is quicker and less demanding in computer storage space but otherwise operates in the same way as Program X with x_4 set to 0.0.

Program XII is a variant of Program X in which the algorithm for eliminating the spectral slit distortion (Program VI) is incorporated. Having accepted the vector of experimental ordinates together with information about the spectral slit function it first corrects these ordinates for spectral slit distortion by pseudodeconvolution with the slit ordinates and then optimizes the x parameters to fit the slit corrected curve; in other respects it is identical with Program X.

In considering the application of the band fitting programs it must be emphasized that there is an inherent danger in attributing too explicit physical significance to the parameters of the component bands; this is particularly true in the marginal cases where it is found necessary to postulate weak hidden components to take account of slight asymmetry in the experimental band envelopes. These dangers are the greater the less the mathematical sophistication of the user of the programs and one must caution against applying these techniques to band analysis without first acquiring a thorough appreciation of the underlying mathematics (see note 17).

A more specific warning is necessary concerning the choice of the number of component band elements. Commonly the fit will improve progressively as more component bands are added, but only the *minimum* number that will generate a reasonable reproduction of the experimental curve can be countenanced; minor weak components that are not evident to the practiced eye on inspection of the experimental envelope should be treated very circumspectly. Where isosbestic systems of absorption bands are involved, as in thermal or concentration studies of equilibrium systems, or rate studies on reactions where absorbing reactants give rise to absorbing products, band analysis of the type considered here can be profitably combined with factor analysis (see note 18). When supplied with information about the accuracy of the measurements, factor analysis can predict the *maximum* number of components that can be recognized as contributing to the band system [70]. The effective combination of these two techniques of band analysis has been demonstrated by Bulmer and Shurvell [71] from whom a computer program for the factor analysis of isosbestic spectral data can be obtained.

E. Analysis of the Band Fit

When the set of optimized x parameters has been obtained by one of the preceding programs, a vector of the ordinates of the calculated spectrum can be generated by Program XIV or XX (cf. Sec. V.C.1). Program XV can then be used to extract the error curve. Used in a different mode Program XV also serves as a general purpose program to subtract one spectrum from another.

When used for band fit analysis the card input to Program XV will be in the standard $cm^{-1} \times 10/T \times 1000$ format; when used for other purposes it will also accept input as cm^{-1}/A, $cm^{-1}/T \times 100$, $cm^{-1} \times 10/A$, cm^{-1}/T, $cm^{-1} \times 10/T$. Four forms of printed and card output are also available. These are $cm^{-1} \times 10/T \times 1000$, cm^{-1}/A and two special outputs in which the difference is normalized over the range <-1000, +1000> or <+200, +800>; in both the latter the abscissal scale is $cm^{-1} \times 10$. The latter scale is useful when the error is to be computer plotted for qualitative display.

When working with spectra having a high noise level it is convenient to separate the mismatch of the experimental and calculated curves due to the noise element from that due to the basic curve shape. There is an option to do this in Program XV by using a five point quadratic smoothing convolute as a noise filter. Subtraction of the smoothed calculated curve from the smoothed experimental curve gives a measure of the mathematical mismatch after allowance for the noise. The noise spectrum can also be presented separately. This is illustrated in Fig. 8. The program also evaluates and records the root mean square of the ordinate residuals, the algebraic sum of the residuals and the ordinate value and wave number of the maximum absolute residuals of the raw difference curve, the smoothed difference curve, and the noise curve.

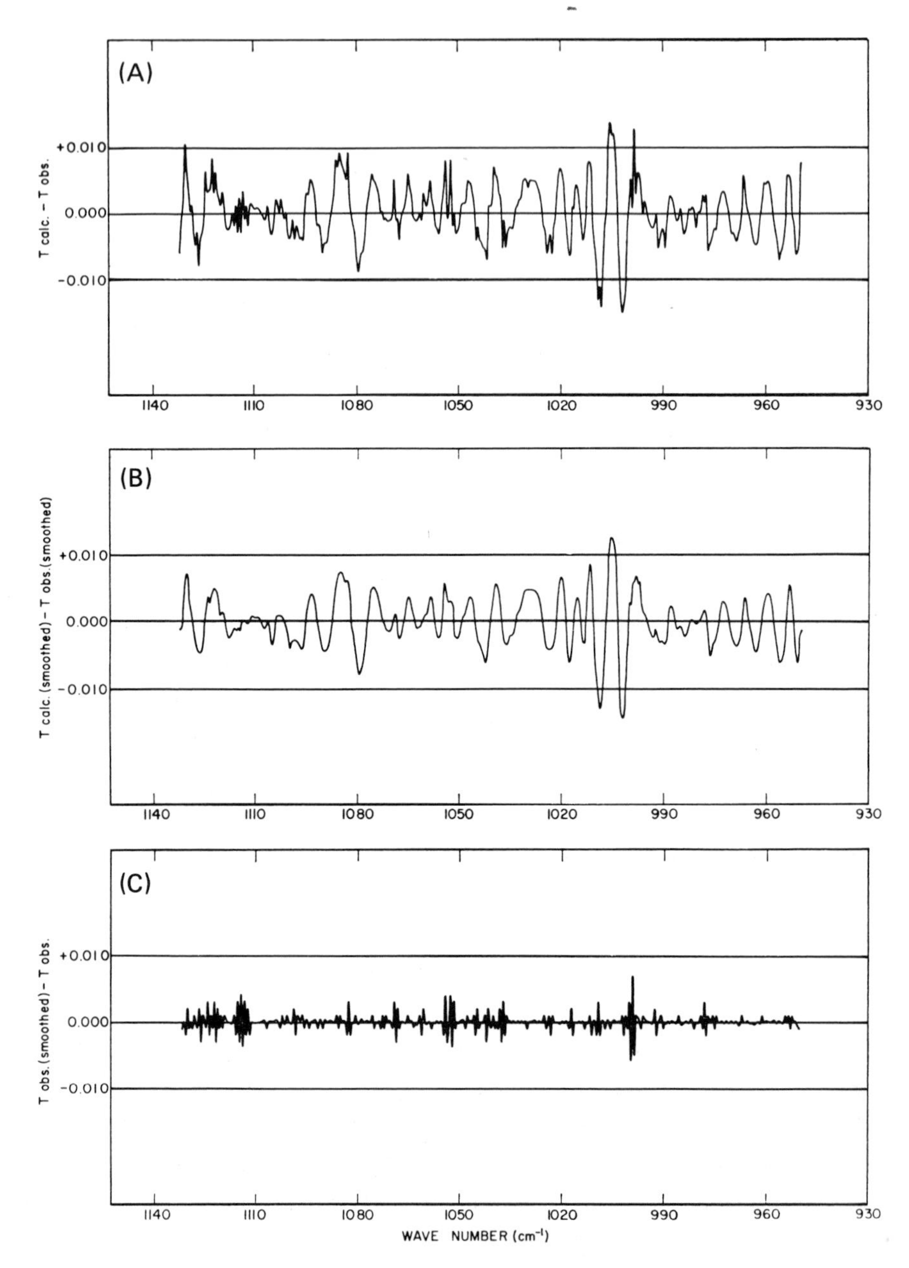

FIG. 8. Program XV. Diagram illustrating separation of the noise error from the band misfit error. Curve A: total error. Curve B: curve misfit error. Curve C: noise error. These curves were obtained in the course of fitting a 16-band section of a steroid spectrum [55].

VI. CALCULATION OF OPTICAL PROPERTIES

Our laboratory has recently been concerned with infrared transmission measurements on pure liquids [14,37-39] and need has arisen during this work to study several optical problems involving reflection and transmission measurements in the infrared and the application of Kramers-Kronig transformations to the evaluation of optical constants from transmission measurements. Programs XLIII, XLIV, XLVI, and XLVII which are described in this section were written in connection with this work.

A. Transmission and Reflection of a Nonabsorbing Plate

Program XLIII computes both the transmittance and the reflectance of a thick free-standing nonabsorbing plate located in the sample compartment of a conventional double-beam spectrophotometer. The thickness of the plate (>5 mm) conforms to the common specifications of infrared cell windows and the interference effects within the plate need not be considered. The algorithm is based on a classical treatment derived from Fresnel's equations for oblique reflection and transmission at an interface; it is discussed in Ref. 38 and in the introductory description to the program [12]. It takes account of the convergence of the incident beam which is integrated over a range of incident angles from zero to the semiangle of the convergent beam (θ_0) using a five-point Gauss quadrature. Account may also be taken of the polarization discrimination of the spectrophotometer but for isotropic plate materials this is a very small effect because of the near-conical symmetry of the incident beam.

The input to the program is the beginning and ending of the desired wave number range, the wave number interval, the beam convergence angle (see note 19), and a vector array of the refractive index of the plate material at a range of wave numbers spanning the computation range; these should be close enough to allow linear interpolation and are obtainable from Program XXXVIII (cf. Sec. IV.E). If a correction is to be applied for instrumental polarization discrimination a set of polarization coefficients as a function of wave number must also be introduced on punched cards. The output is a tabulation of the wave number, the refractive index, the reflectance, and the transmittance of the plate. There is no provision for card output.

B. Transmittance and Reflectance of an Absorbing Film

The experimentally derived expression for absorbance [$\log(P_0/P)$] presupposes that all radiation not transmitted is either absorbed or is compensated by a control cell in the reference beam of the double beam spectrophotometer and that the radiation suffers no multiple reflection within the absorbing layer. In the absence of complete compensation it is more correct to call this measured quantity the *apparent absorbance* (A_{app}); there will be a similar expression k_{app} for the apparent *absorption coefficient* where

$$k_{app} = \frac{2.30258A_{app}}{4\pi\nu d} \tag{61}$$

while from dispersion theory one can compute

$$k_{true} = \frac{2.30258A_{true}}{4\pi\nu d} \tag{62}$$

Program XLIV computes $k(\nu)_{app}$ from $k(\nu)_{true}$ for a thin film between plates of identical window material under conditions where there is no compensation in the reference beam.

One may write formally

$$k_{app(\nu)} = f_1(\hat{n},n_W,\sigma,p)_\nu;f_2(\theta,d,\Delta) \tag{63}$$

where $\hat{n}$ is the complex refractive index of the absorbing material, n_W the refractive index of the window material, σ a diffuse reflectance coefficient determined by the degree of perfection of the window polish, and p a polarization discrimination factor of the spectrophotometer; these terms are all wave number dependent. The quantities grouped in f_2 are θ, the angle of incidence of the radiation, d the mean path length of the cell, and Δ a quantity which determines the nonparallelism of the cell windows; these are geometric quantities independent of the wave number (cf. Fig. 6).

Program XLIV computes $k(\nu)_{app}$ over a defined wave number range in accord with considerations developed in Ref. 38; in so doing it takes account of the internal reflections within the cell and interference effects within the film; n_w can be obtained from Program XXXVIII (cf. Sec. IV.E) while d, σ, and Δ can be obtained by Program XLV (cf. Sec. IV.D.1) from the fringe pattern of the empty cell. The p coefficients can be measured with a polarizer and θ is known from the geometry of the spectrophotometer.

Program XLIV yields the total transmittance (T_t), the baseline transmittance (T_b), and the sample transmittance (T_s) as %T and as optical density ($\log T^{-1}$). The baseline is computed for a nonabsorbing film of the same thickness as the sample but with k = 0 and a refractive index (n) which may be taken as either (i) the averaged refractive index across the band system (constant baseline) or (ii) a refractive index increasing linearly across the band system (dispersive baseline). The apparent absorption coefficient [$A_{app}(\nu)$] is also calculated. The output is a tabulation of these quantities together with the refractive index and related reflectance from the front face of the cell.

C. Determination of Optical Constants from Transmission Measurements on Thin Liquid Films

Equation (63), which forms the basis of the preceding program, mainly has relevance to theoretical studies where the true optical constants are known a priori from other types of measurements such as attenuated total reflection [72,73]. More commonly the inverse problem is encountered in which the $A_{app}(\nu)$ and $k_{app}(\nu)$ spectra are known and it is desired to compute the optical constants $n(\nu)$ and $k_{true}(\nu)$. This can be expressed formally

$$k_{true}(\nu) = F_1(k_{app}, n_w, \sigma, p)_\nu ; F_2(\theta, d, \Delta) \tag{64}$$

and a similar expression for $n(\nu)$. Unfortunately one cannot derive F_1 and F_2 by an algebraic rearrangement of Eq. (63) and it is necessary to adopt an iterative approach.

Program XLVI provides the means of doing this. The algorithm is too complicated to discuss here but it is developed in detail in Ref. 39. We need only note here that a critical step involves the repeated solution of a Kramers-Kronig transform equation

$$n_i = n_r + \frac{2}{\pi}\left[P\int_0^\infty \frac{\nu k(\nu)}{\nu^2 - \nu_i^2}\,d\nu - P\int_0^\infty \frac{\nu k(\nu)}{\nu^2 - \nu_r^2}\,d\nu\right] \tag{65}$$

where n_i is the refractive index at wave number ν_i and n_r is the refractive index at wave number ν_r which is known from a separate experiment. It is desirable, but not essential, that ν_r lie within the wave number range of the computation.

The input to Program XLVII consists of the parameters n_w, p, θ, d, Δ as described above, together with the wave numbers and intensities of the measured spectrum as $cm^{-1} \times 10/T \times 1000$ in the standard card format. A variety of optional controls to monitor the iteration and the "anchor point" refractive index value (ν_r, n_r) are also entered. There is provision to obtain a less exact solution without resort to the extraneous experiment to obtain n_r. The output consists of a tabulation of ν, k_a, k, n, T × 1000, and δk_ν; the latter is the change in k_ν on the last iteration which is a measure of the convergence.

D. Determination of the Refractive Index of a Liquid in a Transparent or Weakly Absorbing Region of the Spectrum

Program XLVII calculates the refractive index of thin liquid films. The method is a modification of that developed by Kagarise and Mayfield [74] in which the spectrum of the liquid is measured in the region of an absorption minimum in a cell having windows of highly refractive material; germanium is usually used. Because of the high refractivity the spectrum is superimposed on a set of strong interference fringes. The program operates in a variety of modes, one of which assumes complete transparency at the minimum while the others take account of the residual absorption. Input to the program is the parameters n_w, p, θ, d, Δ discussed above and the wave numbers and intensity of the measured spectrum as $cm^{-1} \times 10/T \times 1000$. The program resembles Program XLV (cf. Sec. IV.D.2) in using the Powell optimization algorithm to fit the fringe pattern through the region of the minimum and it tabulates n and k at each wave number. A suitable value from this table can be selected for input as ν_r and n_r in Program XLVI.

VII. OTHER PROGRAMS

In connection with studies of the vibrational analysis of complex molecules [75,76] our laboratory has utilized a set of three programs which have been edited by K. G. Kidd [11]. Programs XXXIX, XL, and XLI are coordinated with one another but do not link into the modular series that form the basis of this monograph. Program XXXIX calculates the Cartesian coordinates from the bond angles and bond lengths of a molecule; Program XL is a modification of the GMAT program of Schachtschneider [77] and calculates the G matrix which is part of the input to Program XLI. The latter program is based on the FPERT program of Schachtschneider for the computation of the normal vibrations of a molecule and their refinement by a least squares method to fit the experimental data derived from the infrared and Raman spectra. Bulletin 15 also contains an extensive discussion of the numerical problems that are encountered in the vibrational analysis of complex molecules and the factors that must be taken into account in the choice of the initial force field. These are illustrated by the application of the algorithms to a trivially simple molecule (water) and a complex molecule (cyclopentanone).

Other current work in our laboratory involves precision measurements of single reflection attenuated total reflection spectra and a series of programs for the reduction of these data will be published in a forthcoming N.R.C.C. Bulletin No. 17. These will include Program XLVIII for the calibration of the ATR goniometer and Programs XLIX and L for the computation of the optical constants from the reflectance measurements.

VIII. CONCLUDING REMARKS

This chapter has dealt exclusively with the development work on computer programming for absorption spectrophotometry carried out in our laboratory and no attempt has been made to set this in a proper perspective with respect to the overall state of the art. Reference to the work of other laboratories has only been included when it has had a direct bearing on our own. In concluding I must tender an apology for these omissions but it would have been impracticable to attempt a full coverage of the field in an article of this length.

ACKNOWLEDGMENTS

Most of these programs were written by my coworkers over the past decade and my grateful thanks are due to D. G. Cameron, D. Escolar, H. Fuhrer, J. P. Hawranek, V. B. Kartha, P. J. Krueger, K. G. Kidd, L. Lompa-Krzymien, H. H. Mantsch, P. I. Neelakantan, J. Pitha, K. S. Seshadri, R. Venkataraghavan, and R. P. Young for their essential contributions. To them must be added our technicians Marjory A. MacKenzie and A. Nadeau and also B. R. Weston of Datacap Ltd., and T. E. Bach of the N.R.C.C. Computer Centre.

APPENDIX: Index to Programs

The programs are numbered from I to L in their sequence in the N.R.C.C. Bulletin Nos. 11-17; this corresponds approximately with the chronological sequence of their development. The PC number appended to the program title is part of our internal cataloging code. This index provides a cross reference with page location in the chapter.

Bulletin No. 11		Page
Program I	Scale Conversion Program No. 1 (PC-131)	7
Program II	Scale Conversion Program No. 2 (PC-138)	7
Program III	Location of Maxima and Minima (PC-120)	14

NOTES

1. Requests for information about these publications should be addressed to Dr. R. N. Jones, Division of Chemistry, National Research Council of Canada, 100 Sussex Drive, Ottawa, Canada, K1A OR6. The Bulletins are available at a nominal charge, but we cannot supply card decks of individual programs. The complete set of programs on magnetic tape can be obtained by prior arrangement.

2. More correctly we ought to write dm^{-1} for this unit.

3. The symbolism and terminology used here conforms with the recommendations of the International Union of Pure and Applied Chemistry (IUPAC) [15], except that we use ν for wave number in place of $\tilde{\nu}$, unless the omission of the tilde superscript may cause ambiguity.

4. Assuming a half-bandwidth ($\Delta\nu_{1/2}$) of 5 cm^{-1} and an encoding interval ($\delta\nu$) of 0.5 cm^{-1} this corresponds to a ratio $\Delta\nu_{1/2}/\delta\nu = 10$; and it is this ratio which defines the relative effect of the smoothing function on the high frequency noise and the true band shape. It is also to be noted that the use of this type of function requires that the wave number encoding interval be constant for differentiation but not necessarily so for smoothing.

5. There can be considerable confusion about these signs. For this there are two causes; (i) it is a universal convention for spectroscopists to plot infrared spectra with the high wave number side on the left. Therefore, with respect to the observed curve, when the eye scans it from left to right we observe $dA/d(-\nu)$ for which a peak corresponds to a sign change from - to +. (ii) Within the program the slopes are computed from the transmission curves; an absorption maximum is a transmission minimum so that for an absorption peak $dT/d\nu$ changes from - to + and $dT/d(-\nu)$ from + to -.

6. In Program VIII, which uses the PEAKFIND algorithm in a subroutine, an attempt has been made to automate the choice be-

tween the 9-point and 15-point functions. The 15-point function is used unless its first or final ordinate value (in absorbance) is less than half the peak absorbance; this would indicate that it extends beyond the half-width of the band. This test is acceptable when dealing with well separated bands but doubtful when applied to strongly overlapping bands.

7. The 98.4/0.8/0.8 and the 1/1/1 mixtures are available from the Aldrich Chemical Co., Inc., 940 West St. Paul Avenue, Milwaukee, Wisconsin, U.S.A. 53233 (Catalog No. 19, 167-1 and 19, 148-5, respectively). The 98.4/0.8/0.8 mixture is also available from the Perkin-Elmer Corp., Norwalk, Conn., U.S.A. 06856 (Catalog No. 186-0010).

8. Program XXX should be used for gas phase spectra.

9. This is a particular example of a very general problem in the digitization of instrumental data of all kinds and Program XXXIV and XXXV have potentially wide relevance beyond the field of spectroscopy.

10. Equation (31) is obtained by rearrangement of Eq. (52) of Ref. 24. Note however that in the latter the denominator should read $2Np\nu(dn/d\lambda)$ and not $2np\nu(dn/d\nu)$; also in the preceding Eq. (5) ν should read ν^2.

11. See p. 378 of Ref. 18.

12. A more elegant Cauchy-Gauss combined curve is given by the Voigt function [24,57] which is a convolution of Gauss and Cauchy functions; we have not attempted to generate optimization programs to fit Voigt functions to experimental curves since the simpler sum and product functions seem adequate.

13. The literature of dispersion theory is confused by a lack of uniformity in expressions for $\hat{n}$. Equation (45) conforms with the recent recommendations of IUPAC [15]. In Ref. 59 the form $\hat{n} = n(1 - ik)$ is used, and in Ref. 60 $\hat{n} = n - ik$.

14. We do not attach physical significance to the sign of x_3 or x_4 since, in Eqs. (40)-(44), these parameters occur only in squared terms.

15. This is not true of the damped harmonic oscillator for which neither the $\varepsilon_2(\nu)$ curve nor the $k(\nu)$ curves of Eqs. (50) and (52) are symmetric with respect to ν_o, nor does the $\varepsilon_{2(max)}$ coincide with the k_{max} ordinate.

16. It should be noted that the use of the symmetrical doublet is not required; it was introduced in Ref. 55 for another purpose.

17. This applies even more forcibly to the use of analog methods of curve fitting where the band envelopes can be matched merely by the appropriate adjustment of a set of dials.

18. Also known as principal component analysis.

19. This will usually be known from the optical geometry of the spectrometer.

REFERENCES

1. R. N. Jones, K. S. Seshadri, N. B. W. Jonathan, and J. W. Hopkins, *Can. J. Chem.*, *41*, 750 (1963).

2. R. N. Jones, K. S. Seshadri, and J. W. Hopkins (with the computational assistance of S. D. Baxter, A. Croteau, and B. W. Attfield), *Nat. Res. Council Can. Bull.*, No. 9 (1962).

3. R. N. Jones and A. Nadeau, *Spectrochim. Acta*, *20*, 1175 (1964).

4. R. N. Jones, *Factors Limiting the Precision of Infrared Spectrophotometry*. Report to the Commission on Molecular Structure and Spectroscopy of the International Union of Pure and Applied Chemistry (1964). This document is obtainable at nominal charge from the Depository of Unpublished Data, Canada Institute of Scientific and Technical Information National Research Council of Canada, Ottawa, Canada.

5. R. N. Jones, *J. Japan. Chem.*, *21*, 609 (1967).

6. R. N. Jones, *Pure Appl. Chem.*, *18*, 303 (1969).

7. R. N. Jones, T. E. Bach, H. Fuhrer, V. B. Kartha, J. Pitha, K. S. Seshadri, R. Venkataraghavan, and R. P. Young, *Nat. Res. Council Can. Bull.*, No. 11 (1968); 2nd ed. (1976).

8. R. N. Jones and J. Pitha, *Nat. Res. Council Can. Bull.*, No. 12 (1968); 2nd ed. (1976).

9. R. N. Jones and R. P. Young, *Nat. Res. Council Can. Bull.*, No. 13 (1969); 2nd ed. (1976).

10. D. G. Cameron, D. Escolar, L. Lompa-Krzymien, P. Neelakantan, and R. N. Jones, *Nat. Res. Council Can. Bull.*, No. 14 (1976).

11. H. Fuhrer, V. B. Kartha, K. G. Kidd, P. J. Krueger, and H. H. Mantsch, *Nat. Res. Council Can. Bull.*, No. 15 (1976).

12. J. P. Hawranek, P. Neelakantan, R. P. Young, and R. N. Jones, *Nat. Res. Council Can. Bull.*, No. 16 (In press).

13. D. J. Cameron, D. Escolar, R. P. Young, and R. N. Jones, *Nat. Res. Council Can. Bull.*, No. 17 (in press).

14. R. N. Jones, D. Escolar, J. P. Hawranek, P. Neelakantan, and R. P. Young, *J. Mol. Struct.*, *19*, 21 (1973).

15. *Pure and Applied Chem.*, *21*, 15 (1970); IUPAC Additional Publication, Butterworths, London (1973).

16. K. L. Neilsen, *Methods in Numerical Analysis*, Macmillan, New York, 1956, p. 656.

17. T. Bulmer and H. F. Shurvell, *Can. Spectry.*, *16*, 94 (1971).

18. Maurice G. Kendell and Alan Stuart, *The Advanced Theory of Statistics*, 2nd ed., Vol. 3, Griffin, London, 1967, Chap. 46.

19. A. Savitzky and M. J. E. Golay, *Anal. Chem.*, *36*, 1627 (1964).

20. J. P. Porchet and Hs. H. Günthard *J. Phys. E; Sci. Instr.*, *3*, 261 (1970).

21. R. N. Jones, R. Venkataraghavan, and J. W. Hopkins, *Spectrochim. Acta*, *23A*, 925 (1967).

22. R. N. Jones, R. Venkataraghavan, and J. W. Hopkins, *Spectrochim. Acta*, 941 (1967).

23. A. Savitzky, *Anal. Chem.*, *33*, No. 13, 25A (1961).

24. K. S. Seshadri and R. N. Jones, *Spectrochim. Acta*, *19*, 1013 (1963).

25. *Recommended Consistent Values of the Fundamental Physical Constants, 1973. CODATA Bulletin No. 11* (1973).

26. R. G. Gordon, *J. Chem. Phys.*, *42*, 3658 (1965); *43*, 1307 (1965).

27. H. Shimazu, *J. Chem. Phys.*, *43*, 2453 (1965); *48*, 2494 (1968).

28. S. Bratoẑ, J. Rios, and Y. Guissani, *J. Chem. Phys.*, *52*, 439 (1969).

29. *Pure and Applied Chem.*, *1*, 537 (1961).

30. *Pure and Applied Chem.*, *33*, 607 (1973) and corrigendum.

31. *Tables of Wavenumbers for the Calibration of Infrared Spectrometers*, 2nd ed. (1976), A. R. H. Cole, ed., Pergamon Press.

32. R. N. Jones and A. Nadeau, *Can. J. Spectry.*, *20*, 33 (1975).

33. D. S. McKinney and R. A. Friedel, *J. Opt. Soc. Am.*, *38*, 222 (1948).

34. A. R. Downie, M. C. Magoon, T. Purcell, and Bryce Crawford, Jr., *J. Opt. Soc. Am.*, *43*, 941 (1953).

35. H. C. Burger and P. H. Van Cittert, *Z. Physik*, *79*, 722 (1932); *81*, 428 (1933).

36. W. F. Herget, W. E. Deeds, N. M. Gailar, R. J. Lovell, and A. H. Nielsen, *J. Opt. Soc. Am.*, *52*, 1113 (1962).

37. R. P. Young and R. N. Jones, *Chem. Rev.*, *71*, 219 (1971).

38. J. P. Hawranek, P. Neelakantan, R. P. Young, and R. N. Jones, *Spectrochim. Acta*, *32A*, 75, 85 (1976).

39. J. P. Hawranek and R. N. Jones, *Spectrochim. Acta*, *32A*, 99, 111 (1976).

40. R. P. Young and R. N. Jones, *Can. J. Chem.*, *47*, 3463 (1971).

41. M. J. D. Powell, *Computer J.*, *7*, 155 (1964); *8*, 303 (1965).

42. B. Elden, *Metrologia*, *2*, 71 (1966).

43. M. Herzberger and C. D. Salzberg, *J. Opt. Soc. Am.*, *52*, 420 (1962).

44. W. S. Rodney and I. H. Malitson, *J. Opt. Soc. Am.*, *46*, 956 (1956).

45. I. H. Malitson, *J. Opt. Soc. Am.*, *52*, 1377 (1962).

46. W. S. Rodney, I. H. Malitson, and T. A. King, *J. Opt. Soc. Am.*, *48*, 633 (1968).

47. I. H. Malitson, *J. Opt. Soc. Am.*, *54*, 628 (1964).

48. I. H. Malitson, *Appl. Optics*, *2*, 1103 (1963).

49. W. S. Rodney, *J. Opt. Soc. Am.*, *45*, 987 (1965).

50. L. W. Tilton, E. K. Plyler, and R. E. Stephens, *J. Opt. Soc. Am.*, *40*, 540 (1950).

51. W. S. Rodney and R. F. Spindler, *J. Res. Natl. Bur. Std.*, *51*, 123 (1953).

52. R. E. Stephens, E. K. Plyler, W. S. Rodney, and R. J. Spindler, *J. Opt. Soc. Am.*, *43*, 110 (1953).

53. *International Critical Tables of Numerical Data, Physics, Chemistry and Technology*, Vol. 7, McGraw-Hill, New York, 1930.

54. F. Paschen, *Ann. Physik*, *4*, 299 (1901); *26*, 120 (1908).

55. J. Pitha and R. N. Jones, *Can. J. Chem.*, *44*, 3031 (1966).

56. J. Pitha and R. N. Jones, *Can. J. Chem.*, *45*, 2347 (1967).

57. W. Voigt, *Münch. Ber.*, 603 (1912).

58. G. Andermann, A. Caron, and D. Dows, *J. Opt. Soc. Am.*, *55*, 1210 (1965).

59. J. Fahrenfort, in *Infrared Spectroscopy and Molecular Structure* (M. Davies, ed.), Elsevier, New York, 1963, Chap. 11.

60. A. Hadni, *Essentials of Modern Physics Applied to the Study of the Infrared*, Pergamon Press, London, 1967, Chap. 4.

61. H. Stone, *J. Opt. Soc. Am.*, *52*, 998 (1962).

62. D. Papoušek and J. Plíva, *Coll. Czech. Chem. Commun.*, *30*, 3007 (1965).

63. J. T. Bell and R. E. Biggers, *J. Mol. Spectry.*, *18*, 247 (1965).

64. J. Pitha and R. N. Jones, *Can. Spectry.*, *11*, 14 (1966).

65. M. R. Hestenes and E. Stiefel, *J. Res. Natl. Bur. Std.*, *49*, 409 (1952).

66. K. Levenberg, *Quart. Appl. Math.*, *2*, 164 (1944).

67. D. W. Marquardt, *J. Soc. Ind. Appl. Math.*, *11*, 431 (1963).

68. J. Meiron, *J. Opt. Soc. Am.*, *55*, 1105 (1965).

69. R. Fletcher and M. J. D. Powell, *Computer J.*, *6*, 163 (1963).

70. Z. Z. Hugus and A. A. El-Awady, *J. Phys. Chem.*, *75*, 2954 (1971).

71. J. T. Bulmer and H. F. Shurvell, *J. Phys. Chem.*, *77*, 256 (1973).

72. A. G. Gilby, J. Burr, W. Krueger, and B. L. Crawford, *J. Phys. Chem.*, *70*, 1525 (1966).

73. G. M. Irons and H. W. Thompson, *Proc. Roy. Soc.*, London, *Ser. A.*, *298*, 160 (1967).

74. R. E. Kagarise and J. W. Mayfield, *J. Opt. Soc. Am.*, *48*, 430 (1958).

75. H. Fuhrer, V. B. Kartha, P. J. Krueger, H. H. Mantsch, and R. N. Jones, *Chem. Rev.*, *72*, 439 (1972).

76. V. B. Kartha, H. H. Mantsch, and R. N. Jones, *Can. J. Chem.*, *51*, 1749 (1973).

77. *Vibrational Analysis of Polyatomic Molecules. VI.* FORTRAN IV Programs for Solving the Vibrational Secular Equation and the Least-Squares Refinement of Force Constants. J. H. Schachtschneider, Technical Report No. 57-65, Shell Development Co., Emeryville, Calif. (1964).

Chapter 2

AN ON-LINE MINICOMPUTER SYSTEM FOR INFRARED SPECTROPHOTOMETRY

James S. Mattson*

Carroll A. Smith

Rosenstiel School of Marine and Atmospheric Science
University of Miami
Miami, Florida

---

*Current affiliation: National Oceanic and Atmospheric Administration, Environmental Data Service, Center for Experiment Design and Data Analysis, Washington, D. C.

I. INTRODUCTION

Infrared spectrophotometry has long had the reputation of being only semiquantitative at best. For this reason, some of the attention being lavished on the recently developed Fourier transform infrared instrumentation is derived from long-held misconceptions regarding conventional dispersive instrumentation. Recently Laitinen [1] described infrared spectroscopy as a discipline which had reached a period of senescence, a view which is not shared by the authors. The dispersive instrument manufacturers are moving slowly into the world of today, an era of sophisticated, computer-based data acquisition and reduction. As they begin redesigning their 10-year-old dispersive instrumentation with modern, high-speed detectors and associate them with low-cost, small computers, the field will enjoy a second childhood. This chapter is a description of our attempt to show what can be done by interfacing an on-line minicomputer to a dispersive infrared spectrophotometer.

The addition of digital data acquisition and reduction hardware to an inherently irreproducible instrument may seem like a foolish proposition at first. However, some of the simplest applications of digital data handling can significantly improve the quality of information from even the least expensive infrared instrumentation. The capability to spectrum-average is a valuable technique for improving the signal-to-noise ratio (S/N) in regions of low transmittance. Averaging n spectra together to form one spectrum, provided that the response time of the detector is taken into account, produces an improvement in S/N equal to the square root of n. This technique has been employed to great advantage in Fourier transform infrared spectroscopy, where averages of 100 scans are often employed in order to increase the S/N on a high-resolution spectrum. Additional advantages are available to the spectroscopist with some kind of computing capability, such as slit function deconvolution, mathematical smoothing, and separation of overlapping peaks.

It is always possible that some infrared spectroscopists will contend that the inherent properties of complex infrared spectra and some samples will make the acquisition of digital data a waste of time. Strongly absorbing samples, which require reference beam matching or severe attenuation to produce a satisfactory spectrum, suffer from the inability of many instruments to yield identical results from one day, or even hour, to the next. The drawing of baselines is a difficult task for even the most experienced spectroscopist, when overlapping bands must be used in a quantitative determination. Few infrared spectroscopists will trust the drawing of baselines to their computer, even though the computer's baseline efforts are much more predictable than those of a group of spectroscopists. Complex spectra, such as those of synthetic polymers, do not lend themselves to quantitative analysis of such things as plasticizers, fillers, etc., except where the substrate spectrum has no interfering absorption bands. With such complex spectra, though, it is often possible to obtain a reference sample which does not contain the analyte. The spectrum of the reference material can then be stored as a baseline to be subtracted from spectra of samples containing the analyte. Using a computer, some fraction of this baseline can be subtracted from the spectrum of a subsequent sample. In this manner, the complex baseline is completely removed from consideration, and the spectroscopist can then focus on the difference spectrum for his analysis.

R. N. Jones, of the National Research Council of Canada, and A. Savitzky and R. W. Hannah of Perkin-Elmer Corporation, began developing the use of computer-aided infrared data reduction in the early 1960s. Perkin-Elmer, at the time, offered an abscissa encoder for their 21-series infrared spectrophotometers that, when combined with a retransmitting potentiometer for the percent transmittance data, could be employed as the fundamental components of a digital data acquisition system. Jones' research group at the NRC began acquiring digital data with such hardware on a Perkin-Elmer 621, and published a substantial amount of their software in NRC Bulletins 11-13 in 1968 and 1969 [2-4]. The NRC system is described in Chap. 1 of this volume, so we will not go into it much further than to point out that it is strictly an *off-line* system of infrared data handling, with no real-time information supplied by the computer to the operator. The advantages of an off-line system involve the ability to handle larger amounts of data, higher level programming, and having access to a wide variety of peripheral devices, while an on-line system offers instant turnaround to the user, real-time error detection, and the opportunity to smooth, ratio, plot, etc., while the sample is still in the spectrophotometer. For specialized applications such as kinetic studies or for internal reflection spectroscopy, where the alignment of accessories is critical, an on-line data reduction system may mean the difference between acquiring some data or no data at all.

The on-line system described in this chapter is built around a Perkin-Elmer model 180 spectrophotometer, but it is applicable to nearly all commercial infrared instruments. It should be pointed out, however, that the Perkin-Elmer 180 is particularly well suited for the task at hand. Jones' group at NRC currently uses a model 180 spectrophotometer as input to their off-line infrared data reduction system, as do Savitsky and Hannah at Perkin-Elmer. At this time, the model 180 has emerged as the instrument of choice for digital infrared spectrophotometry, so we have concentrated on the hardware needed to interface the 180 to a minicomputer. The inherent digital design of the 180 makes some of the interfacing simpler, but the concepts are the same for any other instrument outfitted with (i) an abscissa interval marker, and (ii) a retransmitting potentiometer for the ordinate. In fact, in an earlier paper [5] we described an equivalent interface for an inexpensive ($3,500) Hilger-Watts H-1200 infrared spectrophotometer, which is quite adequate for most routine infrared data acquisition. The 180 provides much more flexibility, and when one is spending $10,000-$20,000 on a data acquisition system, it sometimes seems unrealistic to hang that equipment on a $3,500 spectrometer. However, as with any laboratory instrumentation, the needs of the laboratory must be taken into consideration when considering the acquisition of such an analytical system, and the $3,500 spectrometer may fit your needs, while your neighbor may need a $35,000 machine.

The investment required for the interface between an infrared spectrophotometer and the I/O bus of a minicomputer is surprisingly small. In the case of the Hilger-Watts H-1200, the total cost of the electronic components in the interface was under $450; for the 180 interface, the cost was on the order of $2,500. For laboratories that already possess small computers, there is no additional cost beyond that of the interface. For new installations, the cost of computing hardware has been dropping so rapidly that an investment of under $12,000 in 1974 will get a 16K-word minicomputer with a teletype and three tape cassette transports. In subsequent years, the use of microprocessors and cheaper memories will undoubtedly bring the cost down to half of that.

The addition of an on-line computer to an infrared spectrophotometer offers such substantial benefits that it is unfortunate that the instrument manufacturers have not moved in this direction before now, with the exception of the Fourier transform instrument makers. It is reasonable to expect to see some activity in this field by the dispersive spectrophotometer manufacturers soon, and we anticipate that a new resurgence in dispersive infrared instrumentation will result. The Fourier transform infrared instrument houses have been hammering hard at their ability to spectrum-average, and to dig peaks out of noisy, complex baselines, even though all of these operations are made possible simply by the fact that a computer is an inherent part of an FTS instrument. None of these computational advantages is due to the interferometer itself. Once computers become integral parts of dispersive infrared spectrophotometers, some of the capabilities of the FTS instruments will only be on a par with those of dispersive instruments.

II. HARDWARE CONFIGURATION

A. Description of the Perkin-Elmer Model 180

The Perkin-Elmer Model 180 infrared spectrophotometer is basically a multigrating, dual-beam, ratio-recording instrument. The key abscissa element, the grating carousel, is driven by stepping motor pulses to a large frequency cam. As the frequency cam rotates an electromechanical encoder mounted on the cam shaft sends signals (via contact closures) to the wave number readout display and chart drive motor. At appropriate intervals, microswitches initiate grating changes at 2000, 1000, 500, and 250 cm^{-1}, and a filter cam triggers the positioning of filters in the recombined beam. A front-panel switch enables the operator to step the monochromator in 0.1 cm^{-1} or 0.01 cm^{-1} intervals. The fastest scanning speed of the 180 is 32 cm^{-1}/sec from 4000 to 2000 cm^{-1}, 16 cm^{-1}/sec from 2000 to 1000, etc. The thermopile detector has a fairly slow response time, however, and the fastest practical scanning speed for digital data acquisition is 5 digital data intervals per second [6,7].

The biggest problem with the model 180 is the electromechanical encoder on the frequency cam. It consists of spring wire "feelers" that rotate on a series of gold-plated annular rings on a small (4 in.) etched circuit board. These wire contacts may have a tendency to bounce when they hit a dust particle, scratch, or other impediment. Thus erroneous abscissa and ordinate data may spew randomly from the standard interface, requiring the user to carefully edit incoming data with a software routine. The routine described later in this chapter was developed over a 6-month period, and it (more or less) effectively handles any spurious data caused by bouncing encoder contacts.

Figure 1 illustrates the optical layout of the P-E 180. Some of its common features are as follows. There is an extra focal point in the source compartment, which can be used if a separate source is desired. The sample and reference beams come to a focus (nominally about 11 mm wide) in the sample compartments, which means that many accessories do not work optimally in the 180 unless they are designed to take advantage of this focal point. The second chopper, in the monochromator housing, is used to discriminate against sample self-emission. Shown in Fig. 1 is a polarizer mount between M-15 and the entrance slits, which holds a circular wire grid polarizer, a necessity for serious infrared work. The grating carousel has space for seven gratings, only four of which are required for the normal range of 4000-180 cm^{-1}. The other three, plus the mirror M-21 and the

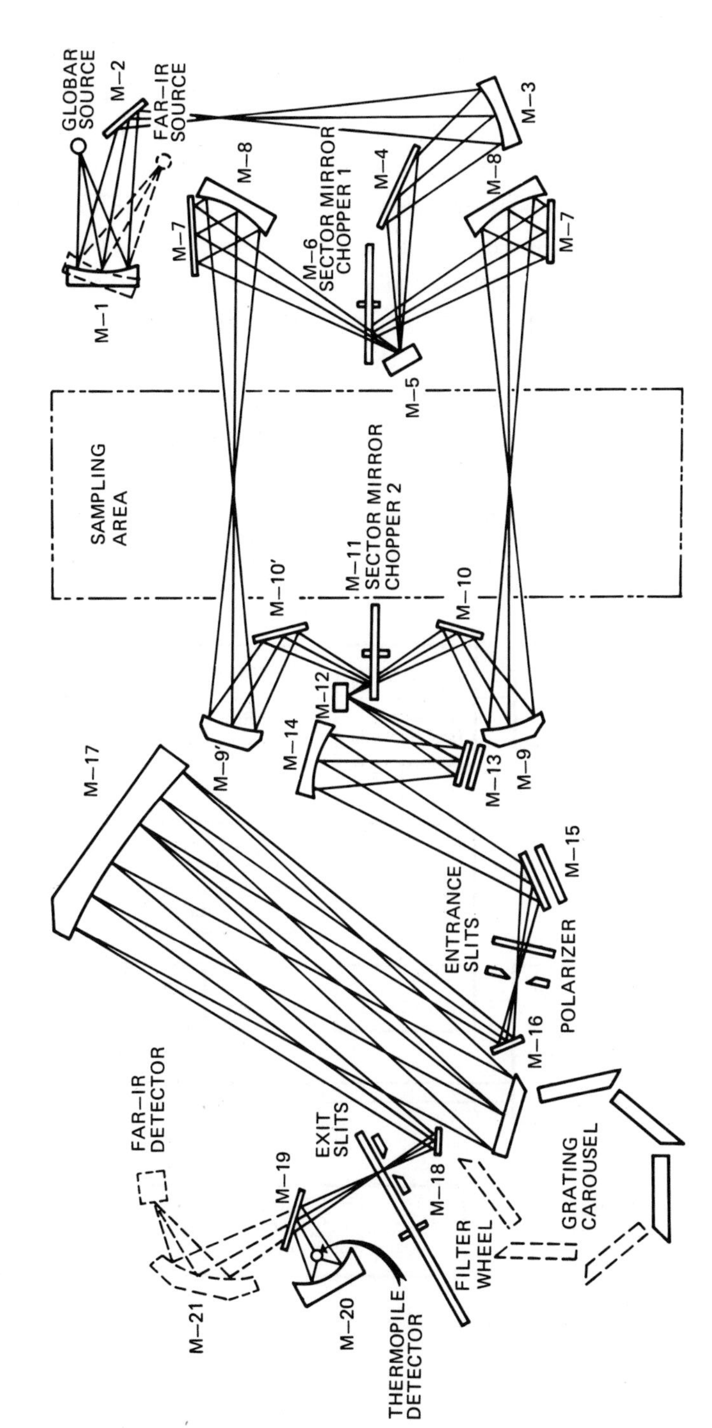

FIG. 1. Optical layout of Perkin-Elmer Model 180 infrared spectrophotometer. (Reprinted by courtesy of Perkin-Elmer Corporation.)

far-IR detector and source, shown in dashed lines, are used to extend the range of the 180 into the far-infrared, to 32 cm^{-1}.

Figure 2 illustrates the model 180 electrical, mechanical, and optical layout in its stand-alone configuration. In the University of Miami system, a DPDT switch installed in the recorder circuitry enables the operator to disable the front panel recorder controls: TIME CONSTANT, LIN A/LIN T, 100%T ADJ, 0%T ADJ, RATIO MODE (I/I_0, I_0/I, etc.), ORDINATE EXPANSION, ABSCISSA EXPANSION, VARIABLE EXPANSION, and SCALE POSITION. In the "COMPUTER" position, this DPDT switch turns over the control of the 180's recorder to the NOVA and the associated software. In the "SPECTROMETER" position, the operator can still acquire digital data with the NOVA, while allowing the spectrophotometer to produce an X-Y plot in its normal stand-alone mode, thus offering the operator a chance to monitor such

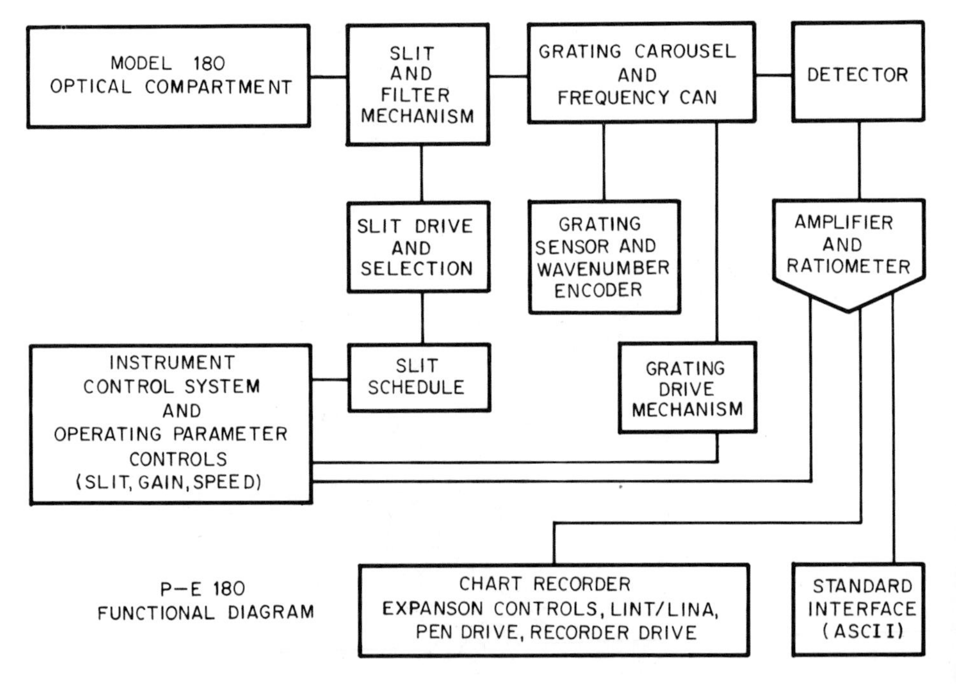

FIG. 2. Block diagram of electronics of Model 180 with standard interface added.

things as the noise level, quality of the dry air purge, level of transmittance, etc., while storing the spectrum in digital format.

B. The Perkin-Elmer Standard Interface

Within the model 180, the ordinate values for T/T_o are converted to ratiometer counts between 1 and 12,000. Normally, the zero is set so that 0%T will produce about 100 to 200 counts, thus providing a "live" zero, yielding about 1 part per hundred resolution in the neighborhood of 0%T. These ratiometer counts must be checked periodically, and are adjusted via the "full count" and "zero count" potentiometers on the ANALOG FILTER pc board in the back of the instrument console. Both zero and 100%T can be reset in the software by appropriate action by the operator, described later, as they are rarely equal to exactly 100 or 10,000, respectively.

The Perkin-Elmer standard interface consists of an additional five printed circuit boards added to the back plane of the spectrophotometer, a data control box, and the necessary connecting cables. It terminates in a 55-pin (Burndy L22TE55SONA-H6 or Cannon KPTMOOE22-55S) female connector. The connecting cable to the computer consists of 15 shielded twisted pairs (Belden 8776), and the connector at the computer is up to the discretion of the individual (i.e., the Cannon DB-25P and S are adequate).

The standard interface transmits both the abscissa and the ordinate, at a switch-selected data interval (0.01 to 10.0 cm^{-1}), in serial ASCII, in the following format:

DATA:	ABSCISSA	< >	ORDINATE	< >	{OPTIONAL}
ASCII:	XXXXXX	<37>	YYYYY	<36>	{<35>}

The first six ASCII characters give the abscissa, most significant character first, to the nearest 0.01 cm^{-1}. The special character (represented by < >) <37> indicates the end of the abscissa ASCII string, and <36> indicates the end of an ordinate string. YYYYY is the number of ratiometer counts, representing I/I_0, between 1 and 12,000. At grating changes, no abscissa or ordinate is transmitted, but the special ASCII character <35> is sent instead. The usual software conversion routines, ASCII to binary, ASCII to floating point, etc., normally terminate an ASCII string and begin conversion on input of a nondigit ASCII character. The special characters <37> and <36> serve this purpose well; the appearance of <35> at each grating change indicates special editing procedures are called for by the data acquisition routine.

The output of the standard interface is TTL logic, 0 and +5 V, and of adequate signal strength and quality to be tied directly to the I/O bus of a NOVA minicomputer. Figure 3 illustrates the digital logic circuit connections at the output connector of the standard interface. Seven shielded twisted pairs transmit the ASCII characters, with BIT 1 the least significant bit. The remaining connections are: (1) the frame ground, (2) the signal ground, (3) the "instrument-on-line" signal, (4) the "data logger-on-line" signal, (5) the "ready," and (6) the "accept" signals. The 180 and the standard interface are extremely sensitive to ground loops, so the frame and signal grounds are very important. In our system, the shields of all of the twisted pairs are tied together and connected to the computer frame and to pin FF, the frame ground of the 180. The 180 signal ground (HH) is connected to the signal ground of the computer via a single, unshielded connection. In the NOVA, frame and signal grounds are the same, so this procedure effectively shorts the 180 frame and signal ground as well, with no apparent problems. It is very important to see that only one electrical ground connection exists in the entire system, including the spectrophotometer, computer, and all peripherals.

The two signals "data logger-on-line" and "instrument-on-line" control panel lights on the data interval box. The "instrument-on-line" signal goes high when the 180 is turned on and the monochromator switch is in the right position, 1:1 for data intervals 1, 2, 5, and 10 cm^{-1}, or 10:1 if the data interval is less than 1 cm^{-1}. The "data logger-on-line" signal must be made true by the computer interface circuitry in order to get the standard interface to transmit data. In our case, it is just shorted (BB to CC).

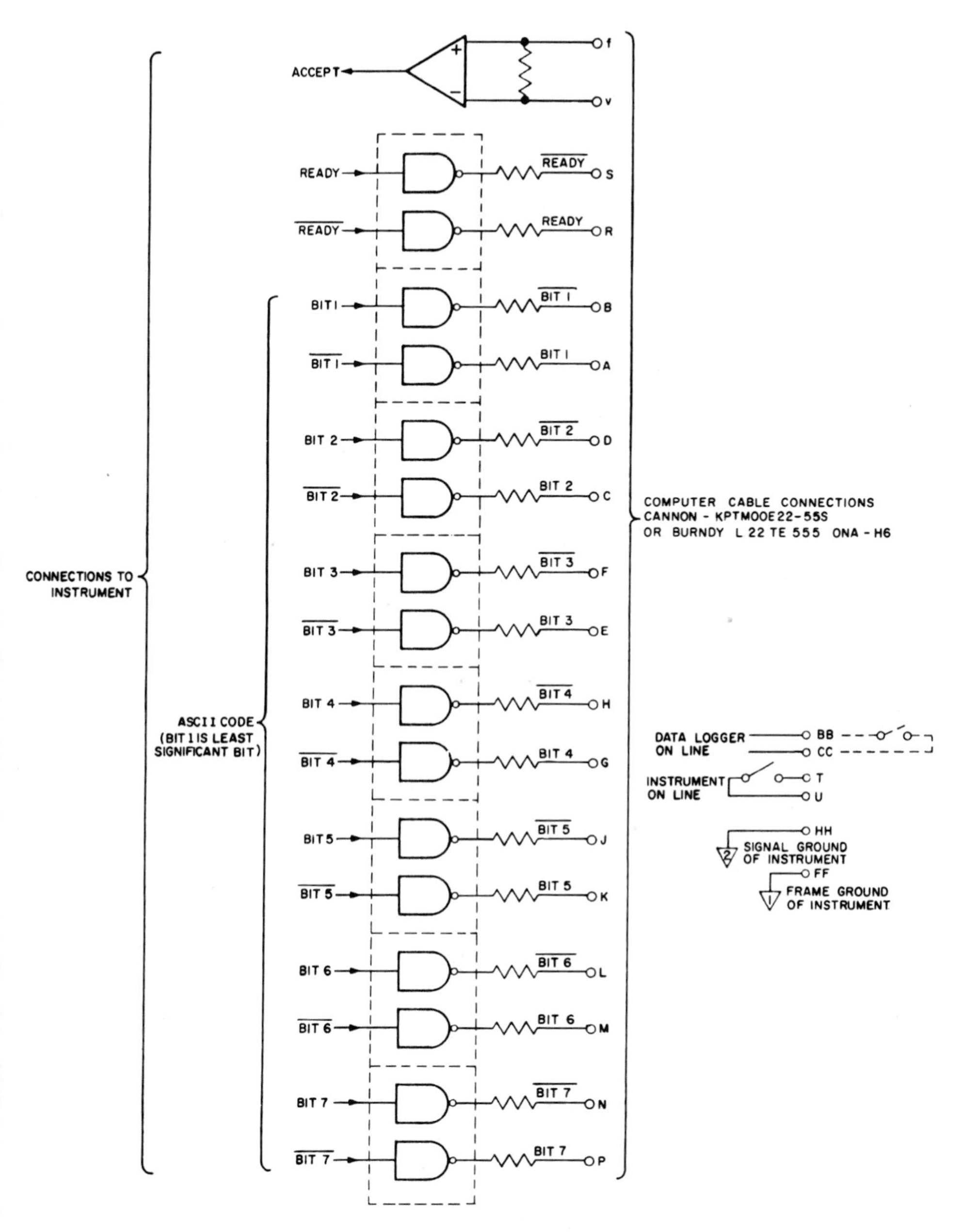

FIG. 3. Digital data connections at output of standard interface (Reprinted by courtesy of Perkin-Elmer Corporation).

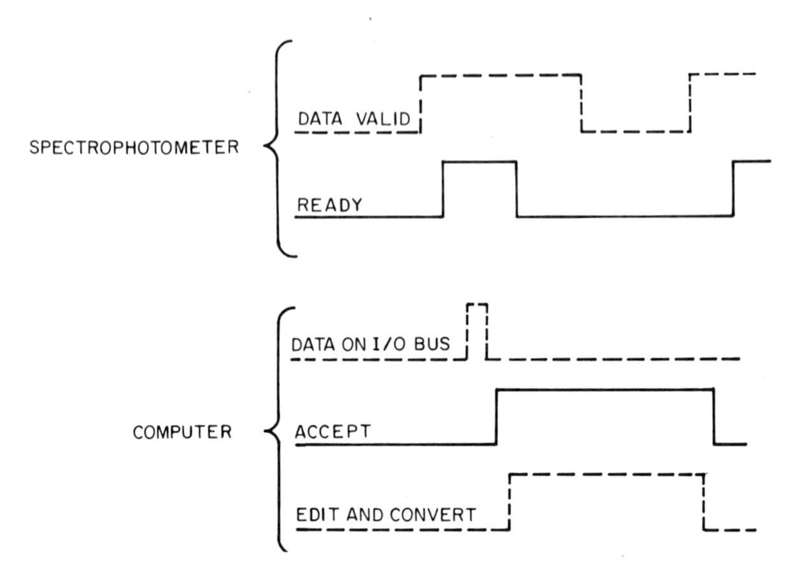

FIG. 4. Timing diagram for digital signals between spectrometer and computer.

The remaining two signals, "ready" and "accept," control the transfer of data to the computer. The ready signal is generated by the standard interface when data is ready on the data lines, and the accept signal is zero. Figure 4 is a timing diagram for this sequence. Once "ready" is high, the computer can strobe the contents of the data lines onto the I/O bus, but it must first set "accept" high in order to reset "ready." After logging the data, the computer resets "accept" to zero, and the interface gets up the next ASCII character. The software sequence is simple, as follows:

1. RESET ACCEPT FF LOW
2. WAIT UNTIL READY IS HIGH
3. TAKE DATA AND SET ACCEPT FF HIGH

C. NOVA Interface

Since the output of the standard interface is serial ASCII, in TTL logic, there is no problem in making the necessary connections to the I/O bus of the NOVA. For a different minicomputer, the interface network will be similar, so the following section is intended to be general enough to allow its application to any external computer.

The NOVA family of computers includes the original NOVA, the NOVA 800 and 1200 series, the Supernova, and the NOVA 2. All of the NOVA series computers use the same basic mechanical design, instructions, and basic architecture. Instruction execution times vary from one to another, from a 2.6 μsec cycle time for the original NOVA processor to 0.8 μsec for the 800 series (0.3 μsec for solid-state memory on the Supernova). The Miami system uses a NOVA 1220, with a 1.2 μsec processor cycle time. The XX20 suffix merely refers to the number of I/O board and memory slots in the main computer housing.

The NOVA has a 16-bit word length and four hardware accumulators, two of which can be used as index registers for indirect addressing. It also has a hardware 15-bit program counter to keep track of the address of the current instruction. Any memory word can be used as storage for instructions, other addresses, or data. Memory words are byte addressable as well, enabling the user to store data, such as ASCII or 8-bit data, as pairs of 8-bit bytes, thus doubling the usual storage capacity.

Input and output are handled on the I/O bus, using either programmed transfers or interrupts. Programmed transfers are adequate for a device like the 180, since it is not able to hold up data while another device is operating. If the 180 were serviced by the interrupt routine, it should be given top priority. We have compromised by turning over total control to the 180 while it is scanning, while driving the cassette tapes and teletype on an interrupt basis (under Data General's SOS).

I/O on the NOVA between a nonstandard external device and the I/O bus is handled with a general purpose interface board (DGC 4040 board). This interface is mounted on a 15 × 15 inch board that has two 100-pin connectors along the rear edge, and is divided in the middle by a double row of 200 wire wrap pins. The front part of the 4040 board is vacant, and ready for the user's logic, with spaces for 65 14- or 16-pin integrated circuits. The rear part of the board contains the DGC logic printed circuitry and the rudimentary ICs needed for communication via the I/O bus. Of the 200 back-plane connections, available as wire wrap connections on the back-plane of the NOVA processor or at the 200 wire wrap pins in the center of the 4040 board, 48 are unused by DGC's logic and therefore can be used by the device.

The prewired logic on the 4040 board includes the circuitry that connects the interface to the data lines and I/O transfer signals on the I/O bus, logic networks for passing the interrupt and data channel request signals through the interface, and a device selection logic network that allows the interface to recognize its device code. The device code is set by the user by putting in the appropriate jumpers. Prewired I/O transfer signals include BUSY, DONE, START, CLEAR, (special) I/O PULSE, DATA IN A, B, or C, DATA OUT A, B or C, INTERRUPT REQUEST, and INTERRUPT DISABLE.

No logic elements need to be installed in the DGC part of the 4040 board. The input logic is shown in Fig. 5. In Fig. 5, the data lines of the I/O bus are labeled $\overline{\text{DATA XX}}$. Data can be strobed onto the I/O bus only by issuing the software command DIA, followed by the appropriate device code. The logic "1" coming from the OR gate at the right of Fig. 5 triggers the NAND gates of $\overline{\text{DATA 9}}$ through $\overline{\text{DATA 15}}$, putting the seven-bit ASCII character directly on the bus. The interposition of a buffer register on the interface board, between the input data lines and the I/O bus, is not necessary, but can be done. As shown earlier in the timing diagram, the data lines from the spectrometer are held valid until after the "accept" signal is received. If timing difficulties were to be observed in another minicomputer system, the input buffer register might be necessary. A pair of 7226 4-bit shift registers serves this purpose.

The "ready" and "accept" signal interfacing is shown on Fig. 6. In the middle of Fig. 6, the top two flip-flops constitute the input interfacing logic. DGC includes the DONE flip-flop on the basic

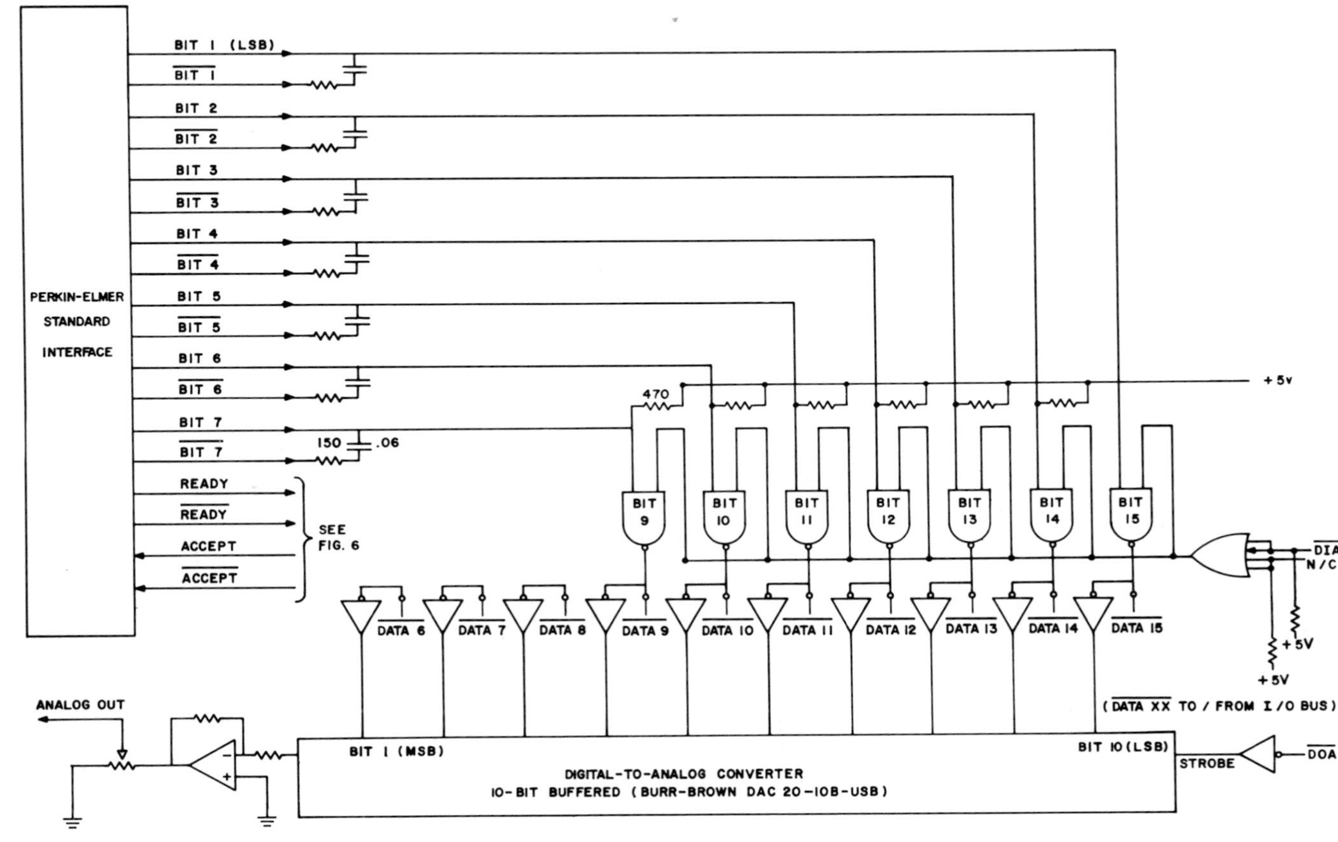

FIG. 5. Input circuitry for ASCII data from spectrometer to I/O bus, and analog output to recorder pen.

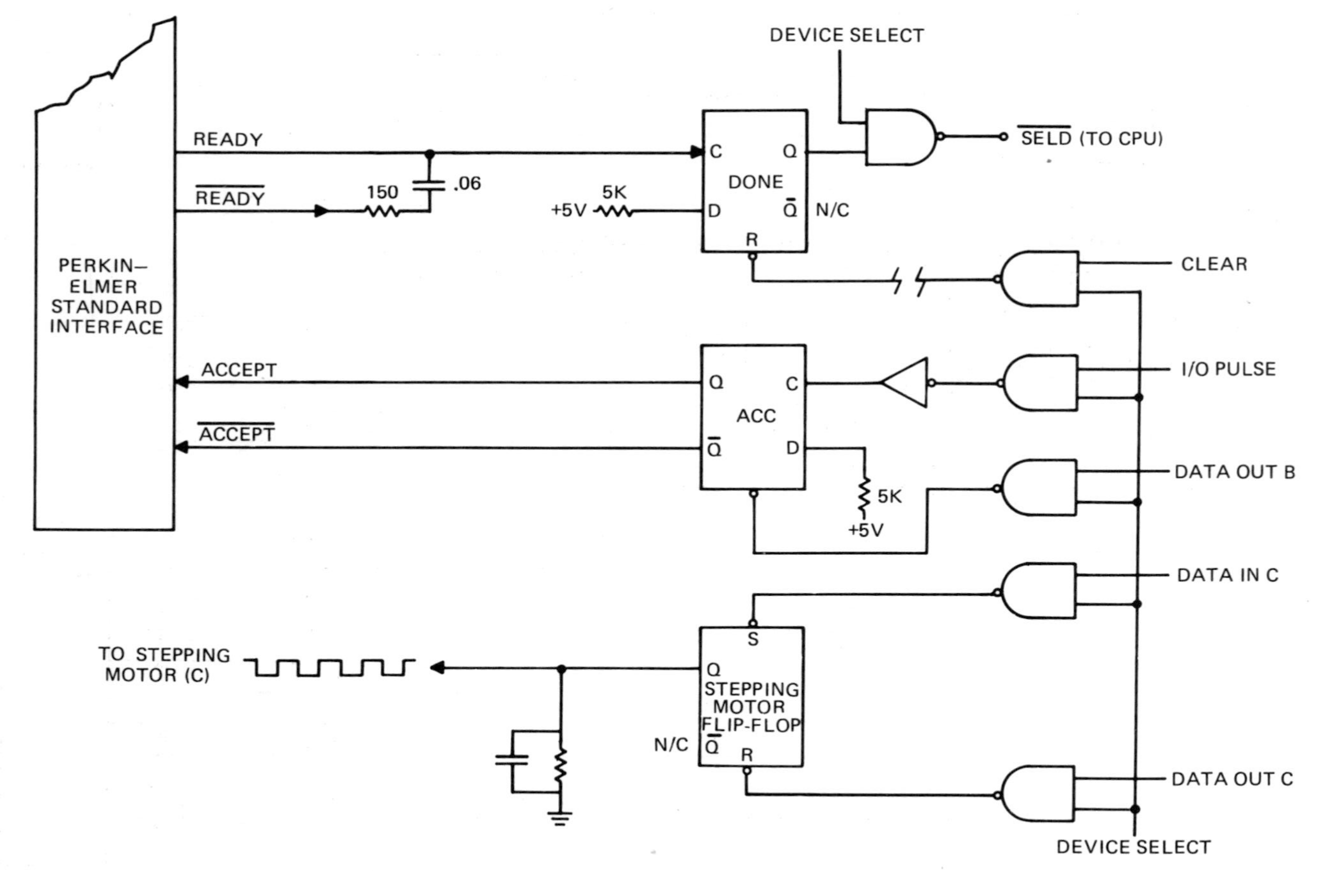

FIG. 6. Ready and accept logic circuitry, and stepping motor pulse generator.

4040 board. This flip-flop is clocked by the spectrometer's "ready" signal, which in combination with a "1" from the device select network (indicating that the software is waiting for data), sets the active state of $\overline{\text{SELD}}$. The software then resets the DONE flip-flop and generates an ACCEPT signal by clocking the ACCEPT flip-flop with the special I/O PULSE. This allows the computer time to take in the data, after which ACCEPT is reset to "Ø" by the action of a DOB command. (Since only one device is connected to this board, the DOB, DOC, DIB, and DIC commands address nonexistent buffers and can be used as extra software commands to the spectrometer.)

The output circuitry is shown in Figs. 5, 6, and 7. Figure 5 illustrates the connection of the digital-to-analog converter to the I/O bus. Binary (scaled) values are continuously varying on the I/O bus, and the DOA signal is used to strobe the contents of $\overline{\text{DATA 6}}$

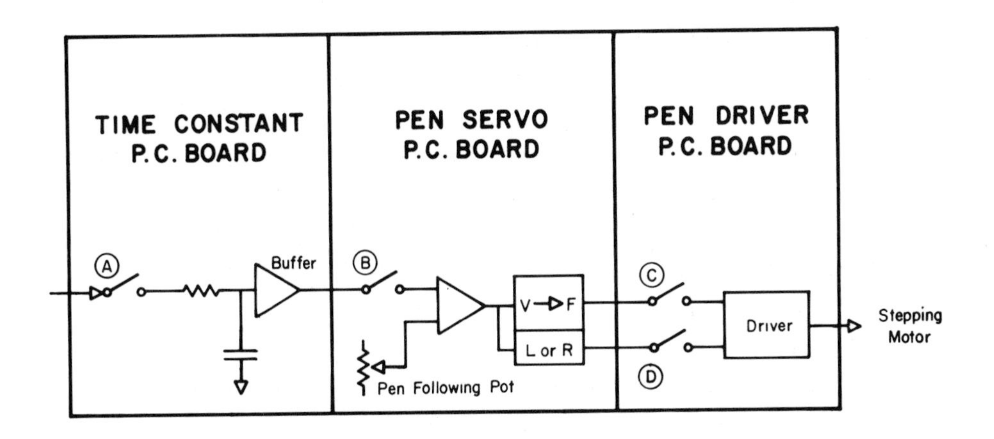

Point A: 0-3.95V Full Scale

Point B: 0-2.37V Full Scale

Point C: 0-0.5V (0-400pps Pulse Rate)

Point D: Motor Direction Control 5v = Right
0v = Left

FIG. 7. Possible inputs to Model 180 X-Y recorder circuitry.

through $\overline{\text{DATA 15}}$ into the buffer of the DAC. Upon conversion, the DAC output (0 to -10 V) is available at the analog output. An operational amplifier is used to invert the signal, and a voltage divider scales it down to 0 to 2.37 V for insertion at Point B on Fig. 7.

On Fig. 6, we show another flip-flop, actually the other half of a dual flip-flop used for the "accept" signal, used as an oscillator. The output of the bottom flip-flop in Fig. 6 is used as a software-controlled pulse generator. Using the $\overline{\text{DIC}}$ and $\overline{\text{DOC}}$ commands to set and reset the flip-flop, the pulse rate is adjustable by software control, and is directly connected to Point C on Fig. 7, the stepping motor driver. The output of the motor pulse flip-flop needs to be fine-tuned by the RC network shown in Fig. 6.

Figure 7 illustrates the four possible inputs from the computer to the chart and pen drive circuitry of the 180. Point A allows the operator to retain the function of the spectrometer's RC pen damping control. Our system uses Point B for the input of ordinate values from the DAC. Two trim-pots connected external to the DAC (not shown in Fig. 5) adjust the full-scale gain and zero offset for the ordinate. The motor pulse output from the interface goes directly to Point C on the pen driver board, where it drives the stepping motor (abscissa control) at a rate determined by a counter in the software. Not shown in the output hardware diagrams are the two power supplies for the DAC, a ±15 V, 300 mA power supply and a +5 V, 500 mA supply.

Table 1 lists the hardware components of the interface. The only major budget items are the "Standard Interface" ($1,500), the 4040 board ($450), the DAC (about $145), and the two power supplies (about $135). The remaining components add up to somewhere around $150, making the total material bill somewhat under $2,500. The time required for assembly is considerably less than that spent waiting for delivery of components, but should not require more than three days once everything is in hand. During checkout, the biggest problem is adjusting the RC network on the analog output of the DAC

TABLE 1

Interface Parts List: 180-NOVA

1	Perkin-Elmer standard interface
1	DGC 4040 interface board
1	Connector, male, 55 pin, Burndy L22T55P6NA-H64
10 ft.	Wire, 15 twisted pairs, Belden 8776
1 ea.	Connectors, male and female, Cannon DB-25P and S
1 ea.	Connectors, male and female, Cannon DB-9P and S
1	Digital-to-analog converter, 10 bit, buffered (Burr-Brown DAC 20-10B-USB)
1	Power supply, ±15 V, 300 mA (Sorenson MMD15-.300)
1	Power supply, +5 V, 500 mA (Sorenson MMS5-.500)
1	Dual flip-flop (7474)
1	Hex inverter (7404)
1	Operational amplifier (LM101A)
-	Assorted resistors and capacitors
1	10K Trimpot (zero offset on DAC)
1	2.5K Trimpot (full scale adj. on DAC)
1	DPDT switch
-	Wire-wrap wire, wire-wrap tool, etc.
	(Optional input buffer)
1	Dual OR gate
2	4-Bit shift registers (7226)

to take into account the cable length and the vagaries of the 180's stepping motor. The output of the standard interface cannot be tested successfully without a "black box" which will provide the proper "accept" logic, thus slowing down the data transfer rate so that one can look at the signals with a scope.

The entire interface was constructed by individuals with little experience in digital circuitry, and for that reason it is an extremely simple system. More experienced electronic engineers will undoubtedly find considerable fault with our circuitry, and improved versions will result. The interface described above does satisfy the basic criterion for our application, however, in that it works.

D. Computer-Spectrometer System

Figure 8 is a block diagram of the complete computer-spectrometer system at the University of Miami. The computer and its peripherals represented a capital investment of about $16,000, but newer computers (NOVA 2, for instance) are available at substantially lower cost. A summary of the major capital expenditures for a NOVA 2 system, like that shown in Fig. 8, is as follows (based on list prices as of January, 1974):

NOVA 2/4 central processor with 16,384 words of core	$ 5,600
Interfaces for TTY and cassettes	1,850
Model 33 ASR Teletype	1,250
Three cassette drives	3,500
TOTAL	$12,200

The standard cassettes for a NOVA system will hold about 55,000 16-bit words of data. However, even if one operates the system under DGC's Stand-alone Operating System (SOS) with the writing of interrecord gaps under a predetermined format, the practical capacity of a cassette is about 44,000 words, or 22,000 floating point numbers. Storing spectra as ordinate values only, in floating point

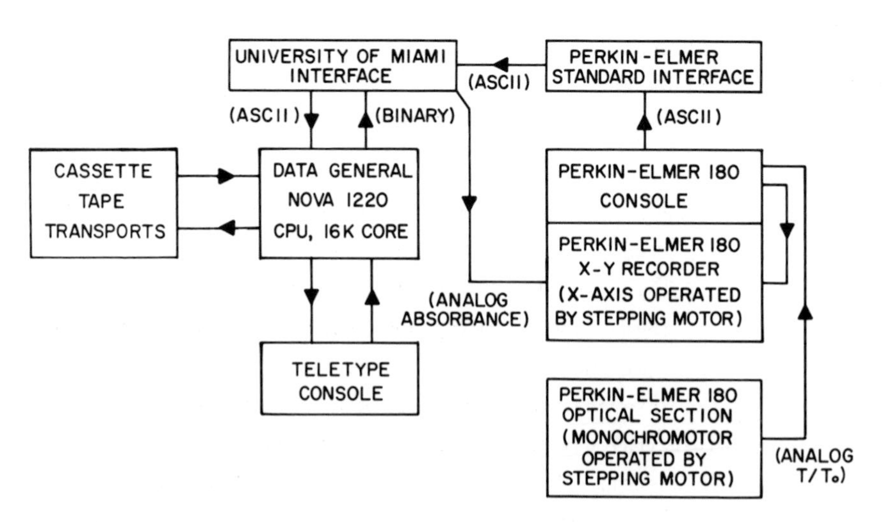

FIG. 8. Block diagram of computer-spectrometer system at the University of Miami.

number format, preceding each spectrum with an ASCII title line, one can store five 4096 point spectra per cassette.

Other peripheral devices, such as large-format magnetic tape, fixed or moving head discs, or even the new "floppy" discs, may be preferable to cassettes. The capital investment required, though, is usually greater than that for cassettes. One could also present a good argument for a CRT-style terminal and a high-speed paper tape reader instead of a standard teletype, but the difference in cost is substantial.

III. INFRARED DATA HANDLING AND REDUCTION

A. Organization of the Main Program "RSMAS"

The main program used to control the University of Miami's infrared spectrometer-computer system is written in Data General's Assembler language, and assembled in relocatable binary format. The main program carries out seven principal functions, selected by the operator in response to the query: *FUNCTION? Using ↓ to represent a carriage return, the seven allowed responses are described below.

P↓ Enables the operator to reset the zero and 100%T counters from their default values of 100 and 10,000 or from previously set values.

C↓ Reads floating point absorbance data from a file on a cassette tape, after echoing the ASCII title line making up the first record of the file.

W↓ Write a data file and an operator-written title line onto a cassette tape.

S↓ Input up to 4096 ASCII data points from the spectrophotometer, to be stored in memory as a string of floating-point ordinate values. The operator can choose to average any number of spectra under this mode of operation, combining them all into one spectrum.

T↓ Type the ordinate values from the spectrum stored in core onto the teletype.

R↓ Ratio two spectra stored as cassette tape files, by subtracting absorbances (scale factor for subtrahend included).

M↓ A response which indicates the operator's wish to either plot or smooth up to 4096 points stored in memory. An M↓ response is followed by the query *SMOOTH OR PLOT?, requesting an operator response of S↓ or P↓. S↓ results in the application of a 21-point quartic smooth to the stored spectrum. P↓ causes the stored spectrum to be plotted on the spectrometer's X-Y recorder.

The main program, the individual sections of which are discussed below, requires only 178 page zero relocatable and 1409 normally relocatable 16-bit words of core memory, plus an additional 8192 words for storage of a 4096 point spectrum (as floating point numbers, each requiring 2 words). The use of floating point numbers requires loading the Floating Point Interpreter tape, a relocatable binary set of subroutines which occupies an additional 1960 core locations. The program runs under the control of DGC's Stand-alone Operating System (SOS), which handles the cassette tape drives and the teletype on an interrupt-driven basis. The set of programs thus required for operation includes the cassette driver, the teletype driver, the SOS library, the Floating Point Interpreter, and the main program, RSMAS. These programs occupy 222 of the 256 page zero locations and 14,131 of the total 16,384 core locations (8192 assigned to spectrum storage). The machine language assembled programs are stored as a core image ("savefile") on a single cassette file, and can be loaded in about 30 sec by the core image loader. (The core image loader resides permanently in high core and is not destroyed by any of the programs.)

In addition to the operations carried out by the main program, individual users have written data reduction programs, including peak locating, integrating, peak separation, slit function deconvolution, etc., in BASIC. DGC's Extended BASIC is also loaded from a cassette savefile in a matter of seconds, and the individual's programs can be loaded under SOS+Extended BASIC by the BASIC keyboard command LOAD "CTX:YY"↓. Individual programs which overrun the capacity of the 16K memory can be linked and swapped using the BASIC "CHAIN" instruction. Thus, one can run programs of any length under SOS+Extended BASIC, even though the SOS library and the BASIC interpreter take up nearly 11,500 words of core. Data files written by the main program are in floating point format, with the first record in each file containing an ASCII title line. Both the title line and the data file are compatible with BASIC I/O logic. The combination of the smaller, more efficient main program, with the higher level programming capabilities of DGC's Extended BASIC, makes the University of Miami computer-spectrometer system much more powerful than one might initially consider possible with a 16K minicomputer.

B. Detailed Descriptions of Main Program Functions

1. P↓. Adjusting Live Zero and 100%T Counters

The Model 180 uses an electronic counter, which counts from zero to as high as 12,000, to represent the ratio of the sample beam intensity to the reference beam intensity (I/I_0). When the sample beam shutter is closed, a minimum count, usually between 50 and 200 counts, is produced. The minimum count provides for a "live zero," so that very low transmittances ($I \simeq 0\%T$) can be recorded with about 1% accuracy. The zero count can be manually adjusted via a trimpot on the analog filter pc board. The full count trimpot on the same pc board can be used to adjust the 100%T count close to 10,000, provided that the sample and reference beams have been properly phased. As a rule, neither the live zero count nor the full scale count match their nominal values of 100 and 10,000, respectively. For this reason, it is necessary to have this routine in order to adjust software counters to represent the actual output of the spectrometer. This is the only way in which one can be certain of matching results over long periods of time, over which time the source may be changed, pc boards may be replaced or adjusted, external temperatures may change, and so on.

Figure 9 illustrates the straightforward logic flow employed to average 100 points with the sample shutter closed, then 100 points with both beams empty, to get representative values for the zero and 100%T counters. This routine sets a switch (SW5) which enables it

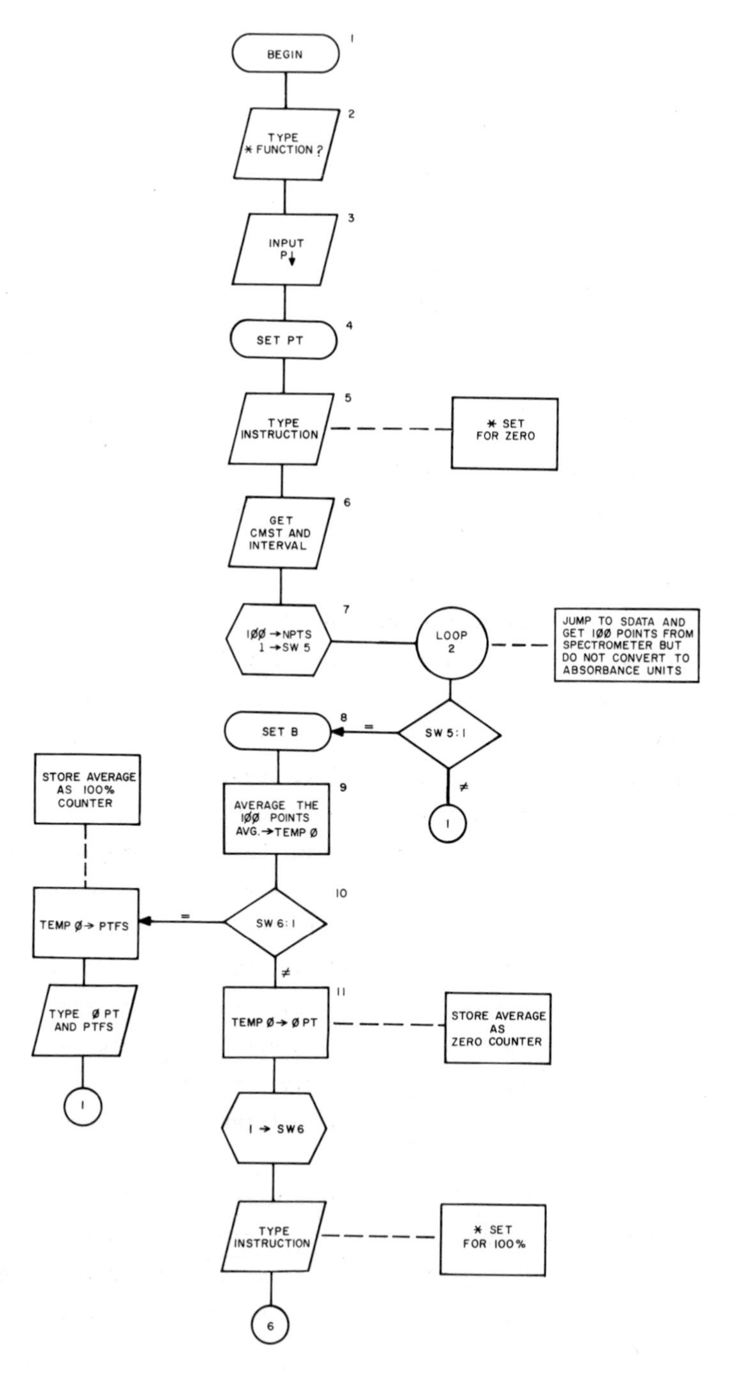

FIG. 9. Flow chart for setting zero and 100%T.

to use the spectrometer data acquisition routine (SDATA), but without modifying the I/I_0 counts in any way.

2. S↓. Spectrophotometer Data Acquisition

A response of S↓ to the query *FUNCTION? generates requests for the starting wave number (CMST), the data interval (INCRE), the number of points to be acquired (NPTS), whether or not averaging is desired, and if so, the number of spectra to be averaged (AVGN). After the computer types *BEGIN, the operator may start the spectrophotometer scan after all of the instrument settings have been made.

According to Savitzky and Hannah [6,7], a model 180 will not accurately respond faster than about five data intervals per second. In addition, our smoothing routine requires about a 21-point halfwidth (FWHM) to avoid distorting the top of a peak. These two restrictions serve to guide the operator in setting the adjustable parameters of the spectrometer. For the narrow absorption bands of, say, a No. 2 fuel oil, a data interval of 0.2 or 0.5 cm^{-1} may be necessary, along with a modest scanning speed (1 to 2.5 cm^{-1}/sec), and a narrow slit width. For broader absorption bands, say those of large proteins, a data interval of 1 or 2 cm^{-1} is sufficient, larger slit widths can be employed, as well as higher scanning speeds. Naturally, the S/N is improved at larger slit openings, thus reducing the need to average spectra in those instances. There does not appear to be any point in scanning at slower speeds than 5 data intervals per second, since no improvement in S/N should result. The S/N is determined by the absolute attenuation of the sample and reference beams, and the gain and slit settings of the spectrophotometer. When the S/N is low, enhancement can be obtained through spectrum averaging and/or smoothing. The operator of the computer-spectrometer system should be familiar enough with the available tradeoffs, both in the spectrometer and in the data handling, to be able to take full advantage of the capabilities of the system.

Figure 10 illustrates the flow chart for the acquisition of abscissa and ordinate data from the spectrophotometer. Using the floating point interpreter (FPI), the ASCII character strings are

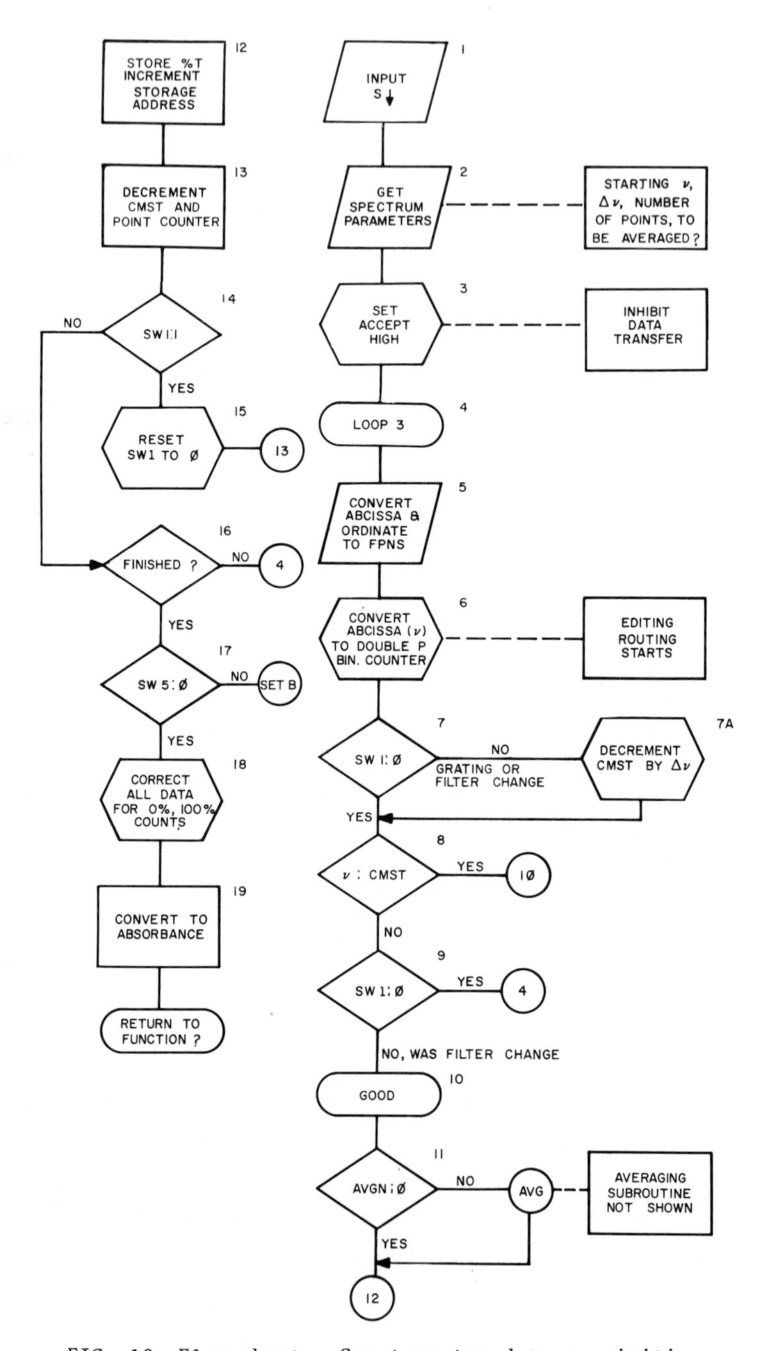

FIG. 10. Flow chart. Spectrometer data acquisition.

converted to floating point numbers. The FPI requires that a get character routine be pointed to by an address in page zero location GETC. The get character subroutine used for data acquisition from the 180 is as follows:

```
GCHAR:  DOB     Ø,Ø5    ;Set accept low
        SKPDN   Ø5      ;Skip next instruction when "done" is set
        JMP     .-1
        DIAP    Ø,Ø5    ;Take character, set accept high
        LDA     2,MSK
        AND     2,Ø     ;Right adjust
        NIOC    Ø5      ;Clear "done"
        JMP     Ø,3     ;Return
MSK:    177
```

Comparing the above subroutine with Figs. 4, 5, and 6 will clarify the actions of the commands "DOB" and "DIAP" (the latter is a combination of "DIA" followed by the special I/O pulse). The appearance of "Ø5" in the I/O commands merely indicates the device code for the particular interface board being addressed.

When the FPI's ASCII-to-floating point conversion routine encounters a nondigit ASCII character, the conversion routine terminates. In the event of a bad conversion, or a nondigit ASCII character that was not preceded by a digit, or an over- or underflow, the FPI provides checkwords for "conversion OK" and "over- or underflow." These checkwords are written in the 100-word writable storage area (WSA) set aside by the main program for use by the FPI. Checking these words, as well as clearing them each time a conversion is called for, must be done by the user. Table 2 is a listing of the assembly language subroutine used by RSMAS to effect these checks.

The single most important feature of the spectrometer data acquisition logic, shown in Fig. 10, is the section which edits incoming data on-the-fly. The editing routine begins at step 8. Basically, it consists of a careful test of the abscissa to determine whether it matches the anticipated abscissa (± 1/10 of the data interval), is a grating change, or is a bad data point. The abscissa is output as a 6-character ASCII string, representing the abscissa to 0.01 cm^{-1}. At a frequency of 3000.00 cm^{-1}, for example, the

TABLE 2

Subroutine to Check ASCII-to-Floating Point Conversion Checkwords

```
CHEK:   STA     3,PSAVØ     ;Save return address
        LDA     2,WSA       ;Get address of FPI area
        LDA     1,Ø,2       ;Get OFLO/UFLO word
        LDA     Ø,1,2       ;Get COK word
        SUBO    3,3         ;Generate a Ø in AC3
        STA     3,Ø,2       ;Clear OFLO/UFLO word
        STA     3,1,2       ;Clear COK word
        MOV     1,1,SNR     ;Test OFLO/UFLO for Ø
        JMP     .+13        ;No OFLO or UFLO
        SUBZL   3,3         ;Generate +1 in AC3
        SUB#    1,3,SZR     ;Test for UFLO
        JMP     .+6         ;Was OFLO
        LDA     Ø,U         ;Was UFLO
        JSR     @PUTC       ;Type "U" on TTY
        JSR     @CRLF       ;CR-LF
        LDA     3,PSAVØ     ;Get return address
        JMP     -4,3        ;Bad datum, get another
        LDA     Ø,0         ;Was OFLO
        JMP     .-5         ;Type "O" on TTY, return
        LDA     3,PSAVØ     ;Get return address
        MOV     Ø,Ø,SZR     ;Test COK for Ø
        JMP     Ø,3         ;Normal return
        JMP     -4,3        ;No conversion, get another
PSAVØ:  Ø
WSA:    .BLK    1ØØ.        ;1ØØ-word storage area
U:      "U                  ;ASCII "U"
0:      "0                  ;ASCII "O"
PUTC:   (Address of subroutine to type single character)
CRLF:   (Address of subroutine to send carriage return-line feed)
```

string 2999.99 cm^{-1} is output, thus each abscissa must be incremented by 0.01 cm^{-1}. The anticipated abscissa, CMST, is compared to the value just received, ν, using double-precision binary subtraction rather than floating point arithmetic to eliminate round-off errors. A precise agreement causes the exit at step 10 to GOOD. Any difference between CMST and ν goes into TEMPØ, where its absolute value is tested against EPSI ($\equiv$ INCRE/1Ø). If ν was within EPSI of CMST, the data are allowed. If not, TEMPØ is rechecked to see if it is greater than CMST, and if so, the data are rejected and a new set of data requested. If not, it is tested to see if ν is only one data interval (±10%) less than CMST. If so, a grating change has just been passed and the data are accepted, setting SW1 in addition. The 180 does not transmit regular data at a grating change, but skips one data interval, thus SW1 is set in order to store the subsequent datum twice. A precaution to the user is to avoid setting up a scan so that a grating change is the last point in the scan. If this is done, the spectrometer must be allowed to pass through the grating change past the next data interval in order to acquire the final point. Other programmers may wish to use the ASCII <35> output from the standard interface as an indicator of a grating change.

As was pointed out earlier, the electromechanical, low-speed, abscissa encoder can transmit substantial numbers of meaningless data if it becomes dirty or scratched. The editing routine described above and shown in Fig. 10 is an absolute necessity to ensure proper registry when plotting, ratioing, or averaging spectra.

The remainder of the flow chart shown in Fig. 10 is straightforward. Once the abscissa is determined to be good, the routine decrements CMST by one data interval and tests AVGN to see if averaging was requested. If averaging is to take place, the subroutine AVG (not shown in Fig. 10) takes the floating point ordinate, divides it by AVGN (the number of spectra being averaged), and adds the result to the previously stored value for the same ν. As the data are entered, they are stored as %T counts (between zero and 12,000 counts) in floating point notation. This enables the computer to take data at very high speeds, much faster, in fact, than the commercial model 180 can scan. At the end of the spectrum (or spectra, if averaging), the entire spectrum is corrected for the live zero, then converted to absolute absorbances (ranging from about -0.05 to about 2 absorbance units).

3. M↓. Plotting or Smoothing Spectrum in Memory

The computer replies to a response of M↓ by requesting the number of points (NPTS), and then asking *SMOOTH OR PLOT? Figure 11 illustrates the logic flow followed in response to an S↓ or a P↓ response. The smoothing routine is based on Savitzky and Golay's [8] moving,

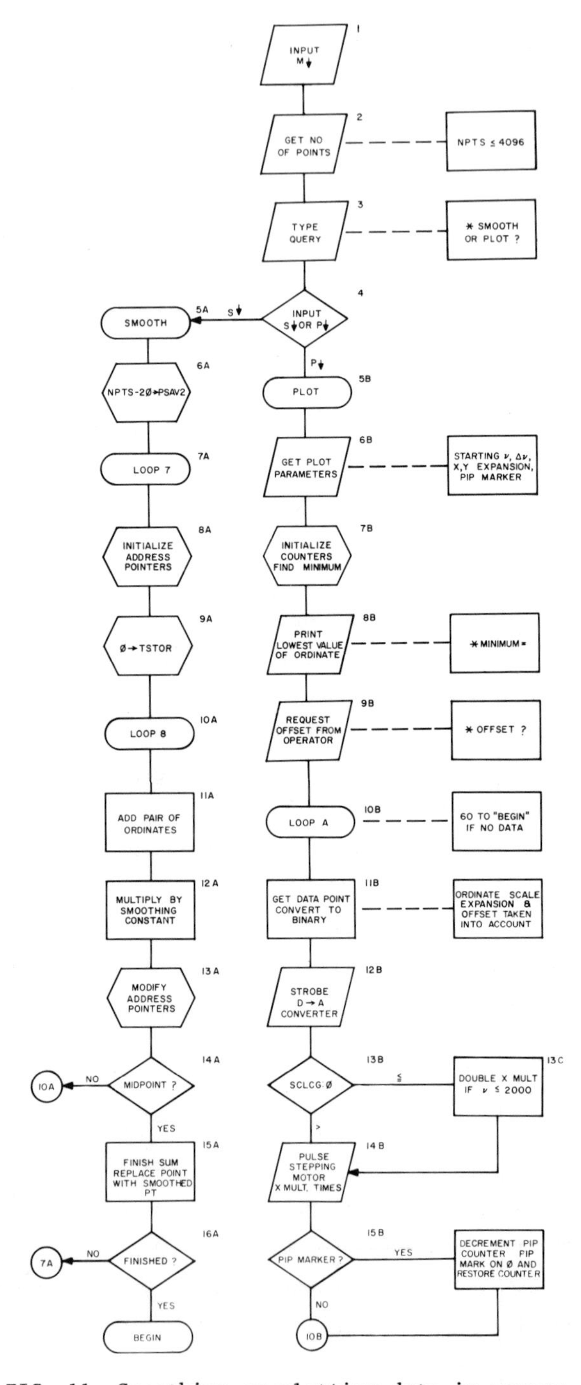

FIG. 11. Smoothing or plotting data in memory.

21-point quartic smooth (as modified [9], and discussed in later papers [6,7]). The 21-point smooth gives a S/N improvement of about a factor of 2 [6,7], and requires that certain precautions be followed when recording the spectrum. First, the half-width (FWHM) of the narrowest peak of interest should not be less than 21 data intervals, otherwise noticeable degradation in the peak height will occur upon smoothing. Second, adjacent peaks which are to remain separated after the smooth must have their peak maxima located at least 21 data intervals apart. When these two conditions cannot be met, the alternative available to the operator is to average four spectra together, with an attendant improvement in S/N of 2 (= $\sqrt{\text{AVGN}}$). In setting up the ν limits, one must remember that, in the 21-point smoothing operation, the first and last 10 points of the spectrum are not modified in any way.

The course of action taken after a P↓ response is shown in Fig. 11. The computer requests sufficient information to set up the plot, including the starting wave number, the data interval, abscissa and ordinate expansion factors (the abscissa expansion factors must be multiples of 1/2 and the ordinate expansion factor may be any positive number from zero to 10^{64}), whether or not a pip marker is desired, and if so, the interval (in points) between pip marks. Next, the computer searches the stored spectrum, finds the minimum absorbance value, prints it, and is prepared to plot the spectrum with the minimum absorbance adjusted to appear as zero on the plotted spectrum. Before plotting, however, the operator is allowed to enter an OFFSET either to displace the spectrum so that the plot will display absolute absorbance values, or to vertically offset the spectrum any amount desired. Negative values for OFFSET should be avoided, since the combination of twos-complement logic in the computer, the single polarity of the D → A converter, and the 0 to 2.37 V range of the recorder all combine to make small negative absorbance values look like large positive values. The computer subtracts 2000 cm^{-1} from the starting frequency in order to set up a binary counter, SCLCG, which will automatically double the number of stepping motor pulses per abscissa when 2000 cm^{-1} is reached on the plot (plotted from high wave numbers to low). Each ordinate, varying between about 0 and 2 absorbance units, is first adjusted for the minimum and offset; i.e., ORDINATE'=ORDINATE-MINIMUM+OFFSET. Since we employ a 10-bit D → A converter, and the unexpanded ordinate scale of the 180 is from 0 to 1.5 absorbance units, each ordinate is multipled by a scale factor of 682.67 before it is converted to a truncated binary integer, and the most significant 10 bits are strobed into the buffer of the DAC. The pip marks are produced by simply adding ten percent of full scale to the contents of the DAC buffer, without moving along the X-axis.

Setting the spectrometer up for a plot requires throwing the DPDT switch (see Fig. 7) to "COMPUTER," putting the pen at the right starting point, setting the monochromator to CONSTANT CM(-1), turning the X-Y recorder ON, and pushing the SCAN button to lower the pen. During the plot, there must be some energy reaching the spectrophotometer detector, or the pen will automatically lift from the paper due to low energy.

4. W↓ and C↓ . Write or Read a Cassette File

Data General's SOS provides the user with simple I/O calls to standard peripheral devices. Calls such as .RDL, .WRL, .RDS, and .WRS, for read a line, write a line, read sequential, and write sequential, make it easy to move data between cassette files and core memory. Figure 12 illustrates both the read and write flow diagrams. Each read or write command requires a byte counter and a byte pointer, which are passed to SOS in the hardware accumulators ACØ and AC1.

Each data file produced by the University of Miami main program begins with an ASCII title line, which is typed in by the operator before the data are transferred to the tape. The title line can contain up to 80 characters (132 are allowed in DGC's default mode), terminated by a carriage return. Pressing "rubout" types a ←, and deletes the last character typed. Pressing "shift-L" causes the entire line to be ignored and a new line requested. The title line should contain such information as the starting wave number, data

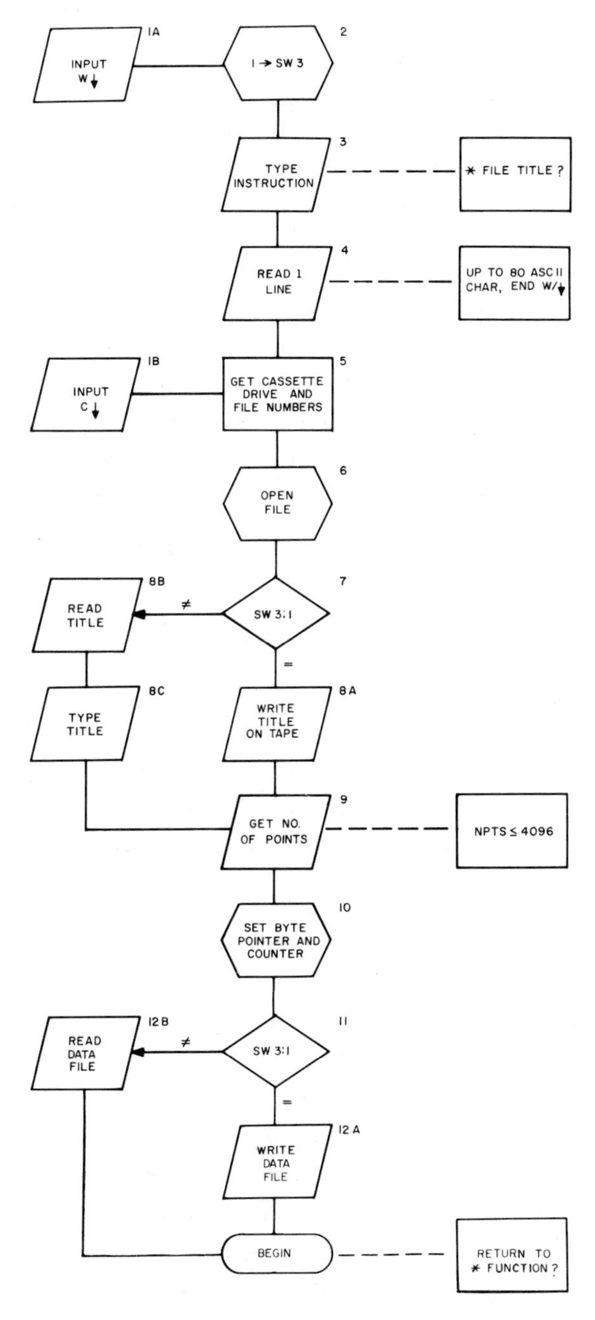

FIG. 12. Write or read a cassette tape file.

interval, number of points, and some kind of sample identification. This makes it easy for multiple users to keep track of their data files over long periods of time.

When a file is to be input following a C↓ response, the title is typed on the teletype before the computer requests additional information from the operator. Data are transferred, both in read and write mode, as a single block of 4·NPTS bytes. Ordinate values are stored sequentially as floating point numbers, each occupying two words. ASCII strings are stored one character to a byte, two bytes to a word. As was mentioned earlier, about 44,000 words can be stored per tape cassette, about five 4096-point spectra and change. One can store six spectra of 3600 points each, or from 4000 to 400 cm^{-1} at 1 cm^{-1} intervals, per cassette tape.

5. R↓. Ratio Two Spectra by Subtracting Absorbances

A flow chart is unnecessary for this simple routine. The spectrum which is the minuend is placed on cassette drive no. 2, and the baseline spectrum goes on cassette drive no. 1. The computer requests the file numbers for the two sets of data, then reads in the entire spectrum from drive 2. A scale factor is then requested from the operator, to be used in a scaled subtraction. Baseline ordinate values are then read in one at a time, multiplied by the scale factor, then subtracted from the minuend already stored in core. The differences replace the spectrum in memory, one point at a time. Since SOS uses a software buffer to transfer bytes in blocks of 510 from a cassette tape, the operation of reading four bytes at a time does not cause jerky tape motion. The above routine does conserve space, though, by not requiring storage for both spectra in core.

6. T↓. Type n Points on Teletype

This is a simple function used primarily for testing purposes. It could be used to produce paper tapes of spectra, but they are not very convenient to handle. To conserve paper, this subroutine simply formats the ordinate data into XXX.XXXXX strings, five to a line, on the teletype. The number of points to be typed are selected by the operator, always started with the first one in memory.

C. Summary of Main Program Routines

TABLE 3

Summary of Features of UM Program RSMAS

Page zero relocatable sections	Word size
Autoincrementing locations	3
Pointers for FPI (GETC, PUTC, WSA)	3
Pointers to normal relocatable sections	44
Text pointers	27
Single-precision binary variables	25
Double-precision binary variables (2)	4
Single-precision binary and ASCII constants	41
Floating-point-number constants (7)	14
Floating-point-number variables (9)	18
Normal relocatable sections and functions	
1. Starting block - type introduction, initialize FPI and SOS.	
2. Initial loop - type FUNCTION, reset all switches, get one of seven valid responses. Return here after each operation.	
3. WDATA and CDATA - write and read cassette files.	
4. SDATA - acquire data from spectrophotometer.	
5. MDATA, SMOOTH, PLOT - as described in Fig. 11.	
6. SETPT - set zero and 100% counters.	
7. RATIO - ratio data from two cassettes.	
8. AVG - averaging subroutine for SDATA.	
9. GCHAR - subroutine to get single ASCII character from 180.	
10. GETTI - subroutine to get single ASCII character from TTY.	
11. PCHAR - subroutine to put single ASCII character on TTY.	
12. TYPIT - subroutine to type message on TTY.	
13. ECHO - subroutine to type data in formatted floating point notation (T↓ response).	
14. CHEK - check ASCII to FPN conversion.	
15. Questions - subroutines to output various queries and instructions, evaluate response, set parameters and jump to appropriate sections.	
16. DADD and DSUB - double-precision binary subroutines.	

TABLE 3 (continued)

Normal relocatable sections and functions
17. Text strings (27 of them).
18. Storage blocks - 100_{10} words for the FPI, 8192_{10} words for one 4096-point spectrum, 40_{10} words for an ASCII title line.

IV. APPLICATIONS OF A COMPUTER-SPECTROMETER SYSTEM

A. Improving Sensitivity and Signal-to-Noise Ratio

There are two simple operations that can be employed to enhance the signal-to-noise (S/N) ratio of dispersive spectrophotometer data, thus improving the sensitivity over that available from the spectrophotometer itself. These operations are spectrum-averaging and mathematical smoothing.

Figure 13 illustrates the improvement obtained in a spectrum after mathematical smoothing, after the procedure described by Savitzky and Golay (Ref. 8, as modified in Ref. 9). In this procedure, a "moving" polynomial smooth is applied to the actual ordinate values, with a degree and span determined by the programmer. The University of Miami spectrum handler utilizes a 21-point, fourth-degree polynomial. In this operation, the eleventh point of the spectrum is forced to conform to its 20 nearest neighbors by the least-squares convention. The smoothed value then replaces the original datum, the 21-point interval is shifted one point to the right, and the next point is smoothed, etc., thus, the term "moving" smooth.

The spectra in Fig. 13 serve to illustrate the improvement in S/N obtained through the 21-point quartic smooth. The bottom line, labeled no. 1, is the difference between the raw data (#3) and the smoothed data (#2) for a fairly noisy, single-scan spectrum of an aqueous solution-germanium internal reflection system.

The process of applying a least squares smoothing function to noisy data is basically similar to the analog concept of an RC filter, except that the analog filter can only use data which have

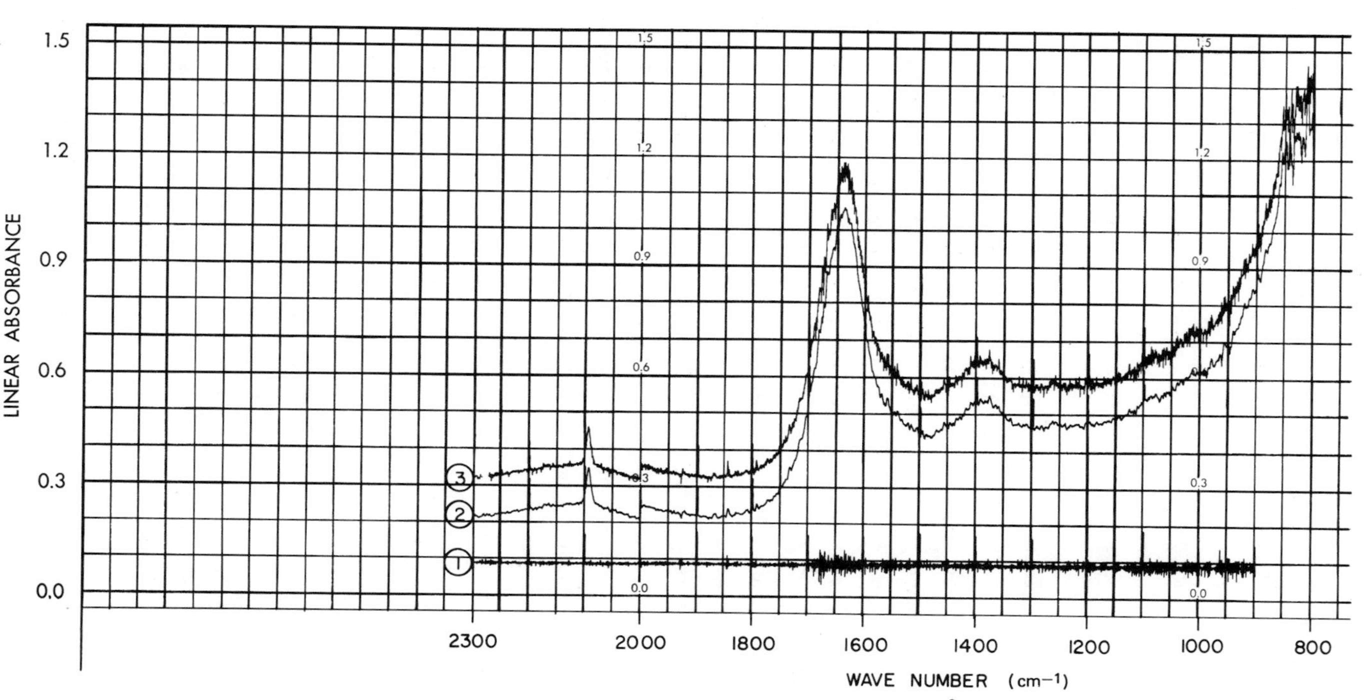

FIG. 13. Example of a 21-point smooth of a spectrum of aqueous CO_3^{2-} and CN^-. Bottom spectrum is difference between raw and smoothed data, showing level of noise reduction obtained.

preceded the present measurement in time, and is thus one-sided. Since the spectrum is already completely stored in the case of mathematical smoothing, the process can use equal intervals of data both before and after the point being smoothed. Therefore, while mathematical smoothing may distort peaks which are narrower than the width employed in the computation, the distortion is one of peak height reduction rather than the distortion of the "right" obtained with RC analog filter networks.

Figure 14 shows that it is not necessary to smooth spectra prior to performing a baseline subtraction. This is valuable to the operator, in that he can leave his "raw" data intact on cassette files, and can smooth the difference spectrum after the subtraction operation is finished. In Fig. 14, the bottom spectrum (no. 1) is the difference between two raw, unsmoothed spectra, showing peaks at 2095 and 1390 cm^{-1} due to aqueous CN^- and CO_3^{2-}. Smoothing the raw spectra and then subtracting produced the difference spectrum labeled no. 2. Spectrum 3, however, is the smoothed version of spectrum 1. There is no difference between spectra 2 and 3, showing that the order of operation is not important. Spectrum 4 is a twice-smoothed version of spectrum 1. It is apparent from spectrum 4 that the second smooth produced no improvement in the data obtained with a single smooth.

Figure 15 illustrates the value of spectrum averaging, showing internal reflection spectra obtained from 1 μg of p, p' - DDT on KRS-5. The system used included a 0.5 mm thick, 25 reflection KRS-5 microprism in a Harrick Scientific microaccessory. One microliter of 1000 ppm p, p' - DDT in hexane (Analabs) was deposited on one face of the microprism, and the solvent was allowed to evaporate. The prism transmitted 18% of the normal energy, at a fixed slit width of 0.7 mm and a resolution of 1 cm^{-1}. Figure 15A shows the raw data obtained from a single, parallel polarized scan from 1150 to 1000 cm^{-1}, at 1X and 10X ordinate scale expansion, unsmoothed and smoothed. The asterisks marks the bands of p, p' - DDT in that region at 1116 (weak), 1094 (strong), and 1016 cm^{-1} (strong). The

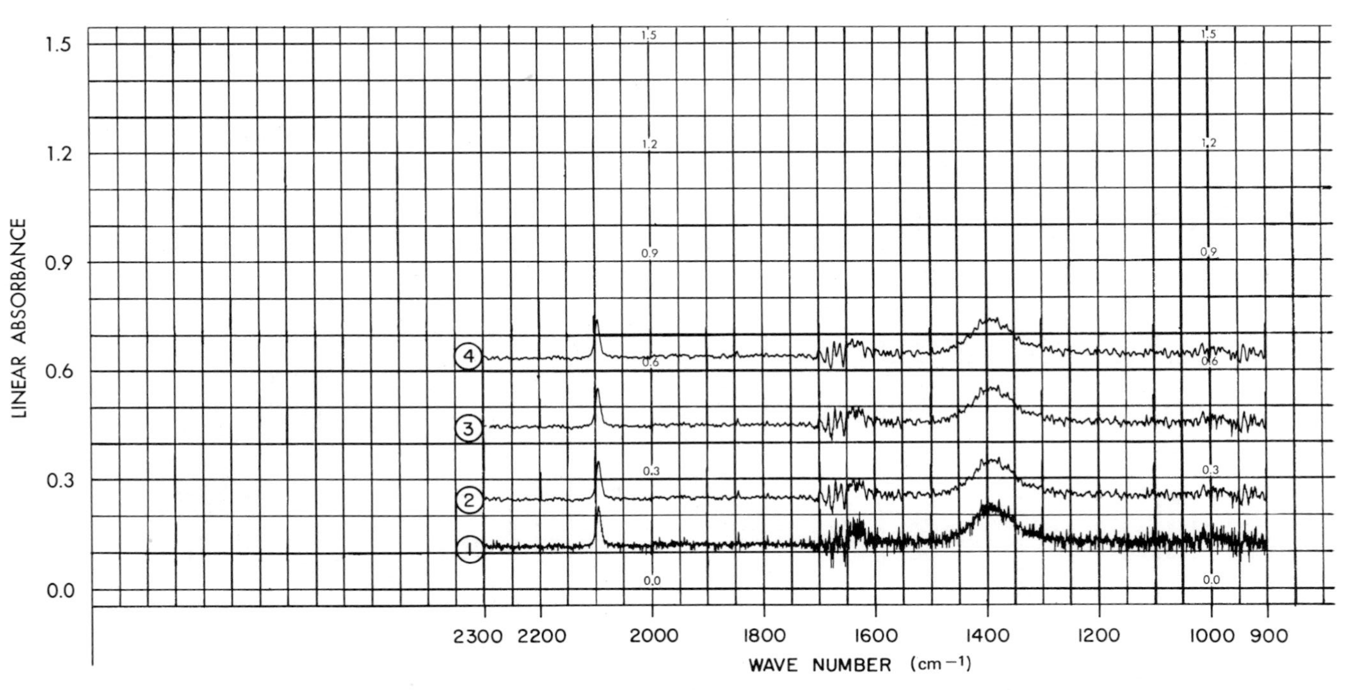

FIG. 14. Showing that the order of smoothing and subtraction does not make any difference. Spectrum 2 is the difference of two smoothed spectra; spectrum 3 is the smoothed difference of two raw spectra.

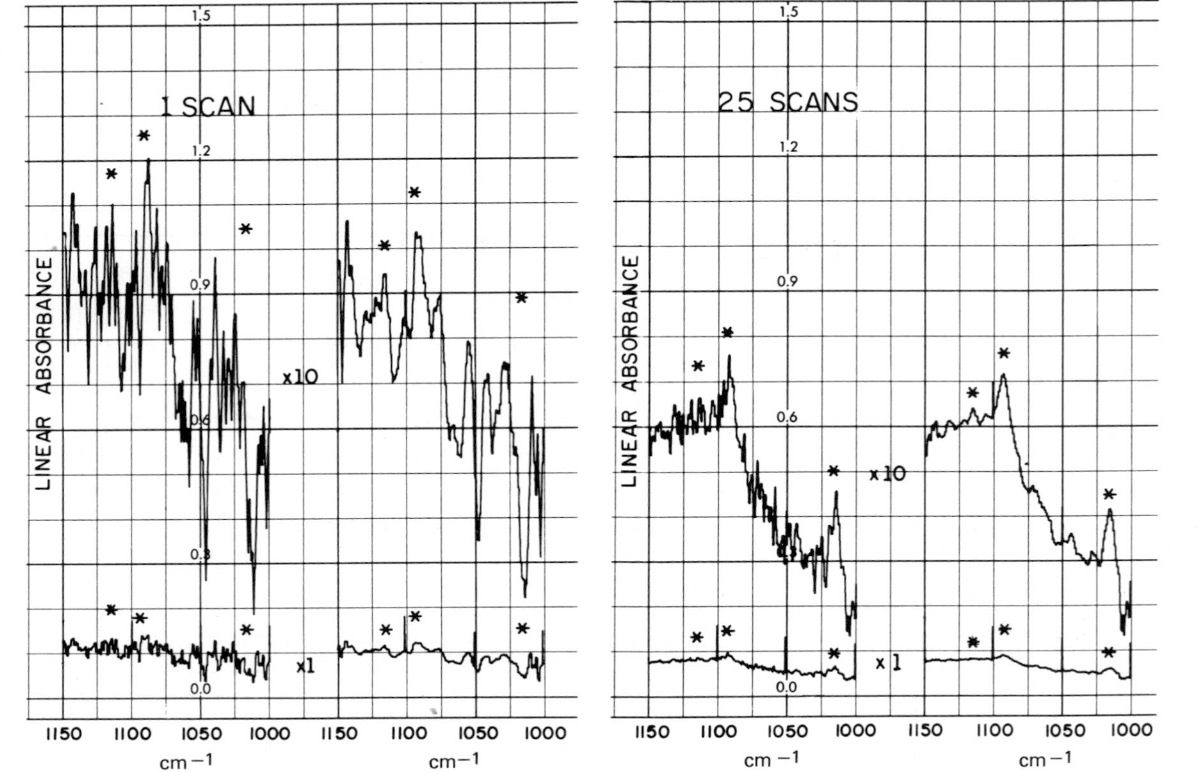

FIG. 15. Signal-to-noise enhancement through a combination of spectrum-averaging and smoothing. (A) illustrates raw (ℓ) and smoothed (r) spectra for a single scan of 1 μg of p,p'-DDT on a KRS-5 micro-ATR prism. (B) illustrates the advantages obtained by averaging 25 spectra of the sample, a process which took less than 25 min, at 4 cm^{-1}/sec. The asterisks indicate the absorption bands expected at 1116, 1094, and 1016 cm^{-1}.

1016 cm^{-1} band does not even show up in the 10X, smoothed spectrum. In Fig. 15B, similar representations are shown for a computer-average of 25 scans. A S/N enhancement of $\sqrt{N}$ is expected for N averaged spectra, and the enhancement is quite evident in Fig. 15B. The strong peaks are easily seen in the 10X spectra, both raw and smoothed and in the smoothed, 10X spectrum some weaker bands are also apparent, including one at 1116 cm^{-1}.

B. Scaled Spectrum Subtraction

One of the most difficult techniques to master in experimental infrared spectroscopy is differential spectroscopy. Differential spectroscopy involves the preparation of a reference cell system which matches the unwanted portion of the sample spectrum perfectly. For instance, aqueous solution spectra, in the transmission mode, require extremely thin (~3 μm) cells of fairly high solute concentrations. Even with such thin cells, the water absorbs so much energy that a compensating cell must be placed in the reference beam, preferably one which can be adjusted until its transmittance matches that of the sample cell. Similar compensating techniques are employed in many other applications in infrared spectroscopy, in order to observe small peaks in the presence of overwhelming solvent bands [10,11].

In. Fig. 14, the difference spectrum between a solution of NaCN and Na_2CO_3 and water was shown. The CN^- and CO_3^{2-} bands at 2095 and 1390 cm^{-1} are not obscured by the strong water bands at 3380 and 1639 cm^{-1}, and there is no difficulty involved in extracting them from the water baseline, as shown in Fig. 14.

Figure 16 illustrates spectrum subtraction in another way, as well as providing some insight into the reproducibility of the spectrophotometer. The familiar spectrum shown in Fig. 16 is actually two independent polystyrene spectra, taken about 2 hr apart, with no averaging or smoothing. The scans were run at 4 cm^{-1}/sec, which is nearly the fastest usable speed for the 180's detector, with normal slit settings and automatic gain control (resolution $\cong$ 8 cm^{-1}

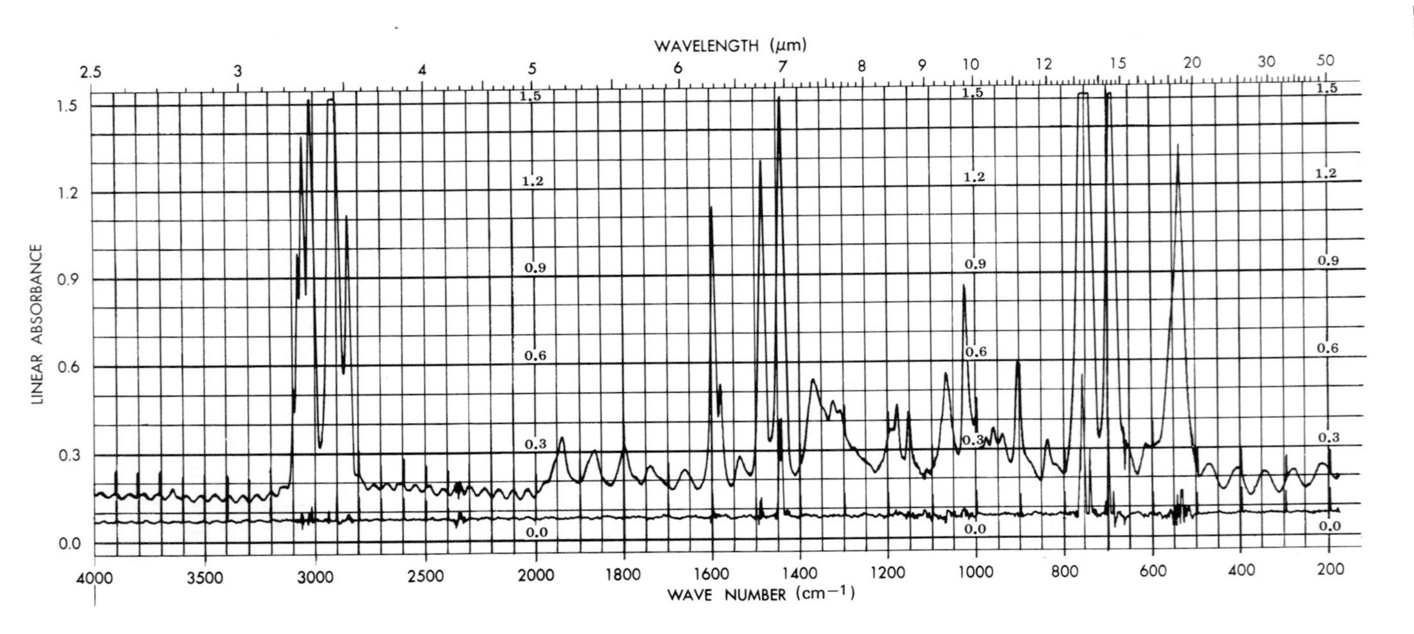

FIG. 16. The polystyrene spectrum shown is really two superimposed scans, taken about an hour apart, at the rate of 4 cm^{-1}/sec. The line along the bottom is the difference between the two individual scans (offset slightly from zero).

at 4000 cm^{-1}). The two superimposed polystyrene spectra were subtracted, with the difference shown at the bottom of Fig. 16. Differences exist at 1437, 754, 737, and 700 cm^{-1}, where the peak absorbances exceed 1.5. The noisier regions at 1200-1000 cm^{-1} and 600-500 cm^{-1} are due to relatively lower energy at the end of those grating ranges. It is reasonable to expect that slightly slower scanning speeds, and possibly averaging or smoothing, would improve the reproducibility of the polystyrene spectra beyond that shown in Fig. 16.

For years infrared spectroscopists have relied on the positions and intensities of the N-H stretching, amide I (primarily C=O stretch), and amide III (C-N stretch + N-H bend) bands as aids in interpreting the secondary structures of synthetic polypeptides and natural proteins. These three absorption bands provide essentially all of the secondary structural information that can be obtained from the infrared spectrum of a polypeptide [12]. Two of the bands however, the N-H stretching band at ~3400 cm^{-1} and the amide I band at ~1650 cm^{-1}, are effectively obscured in aqueous solution spectra by the ~3380 cm^{-1} and 1639 cm^{-1} bands of liquid water. Various efforts to get around this problem have been employed, including the use of D_2O as a solvent, or employing a thin compensating reference cell, as discussed above. The thin compensating cell technique is inadequate for protein solution spectroscopy because of the concomitant requirement of high solute concentrations. D_2O solutions pose an additional problem, that of hydrogen-deuterium exchange. This results in a change in the intensity of the undeuterated amide II band, and formation of a deuterated amide II band at the same frequency as the HDO band [12].

Internal reflection spectroscopy, using a high refractive index, infrared-transparent prism such as germanium, provides the thin sampling region necessary to avoid the total loss of energy at the water peak maxima. For an internal reflection system, the "depth of penetration," d_p, is defined [13] as the distance into the solution where the evanescent field amplitude decays to e^{-1} of its magnitude at the prism-solution interface. For germanium-water, at a 45° angle of incidence, this distance is 0.064 λ_o, where λ_o is the in vacuo wavelength. Of the peak intensity observed in an internal reflection spectrum, 84% is derived from absorbing molecules within one d_p of the prism-solution interface, and 95% of the observed peak intensity is due to molecules within two d_p. Thus for the amide I peak at 1650 cm^{-1}, 95% of the band intensity comes from the 7750 Å thin region at the germanium-solution interface.

Using such a thin sampling region in combination with the signal-to-noise enhancing techniques discussed above, and by subtracting the solvent spectrum from the protein spectrum, the obscured amide I band of a protein can be separated from the overlying water band at 1639 cm^{-1}, as described below.

For solution spectra, four scans at 5 cm^{-1}/sec were computer-averaged for both perpendicular and parallel polarizations. For air-dried films, one scan was sufficient for parallel polarization, two scans were computer-averaged for perpendicular polarization. A precision chopper attenuator (Harrick Scientific), set at 30% transmittance, was in the reference beam, and a twice-normal spectrometer slit program was employed. The cell used to hold the germanium prism is shown in Fig. 17. Intrinsic germanium prisms (Harrick Scientific) employed were 20 × 52.5 × 1 mm oriented single crystals, with the {110} crystal face exposed to the solution. Twenty-two reflections were obtained at the germanium-solution interface.

Protein solutions were prepared from ~72% clottable, ~90% bovine fibrinogen (Sigma), in pH 7.2, 0.1 N phosphate buffer, prepared with doubly distilled, deionized water. The protein solutions were subjected to ultrasonic vibration in order to disperse the fibrinogen. After gathering the solution spectra, the prisms were removed from the cell, rinsed quickly with distilled water, and allowed to air-dry. Film thickness measurements on the dry films were made ellipsometrically by Dr. Robert E. Baier of Calspan Corporation, Buffalo, New York.

The separation of the 1639 cm^{-1} water band from the amide I band of fibrinogen is shown in Fig. 18. The top spectra in Fig. 18

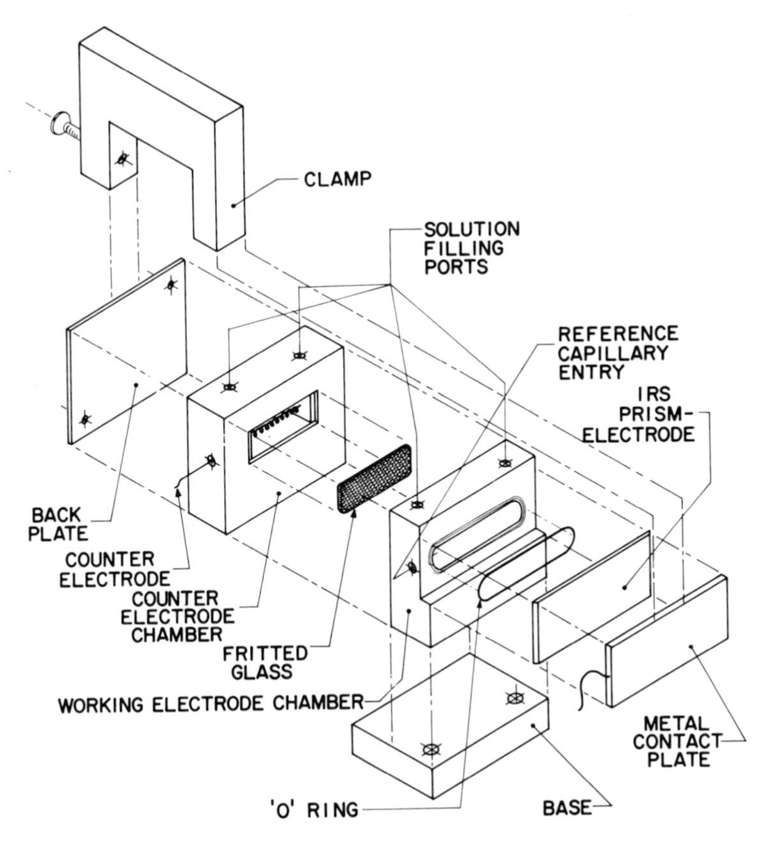

FIG. 17. Spectroelectrochemical cell used for large internal reflection plates.

are averages of four scans of the fibrinogen solution. The amide II band is clearly visible at 1543 cm^{-1} in spectra $1_{\parallel}$ and $1_{\perp}$. The intensity of the amide II peak increased for the first 30 min after the protein solution was added to the cell, indicating a slow spontaneous adsorption of the protein on the germanium surface. Spectra $2_{\parallel}$ and $2_{\perp}$ are the pH 7.2, 0.1 M phosphate buffer baselines, with the $H_2PO_4^-/HPO_4^{2-}$ band appearing at 1070 cm^{-1}. The top four spectra of Fig. 18 are compressed to 0.375X for presentation purposes. Spectra $3_{\parallel}$ and $3_{\perp}$ and the simple absorbance differences of the protein solution spectra less the buffer baselines. The negative water peak

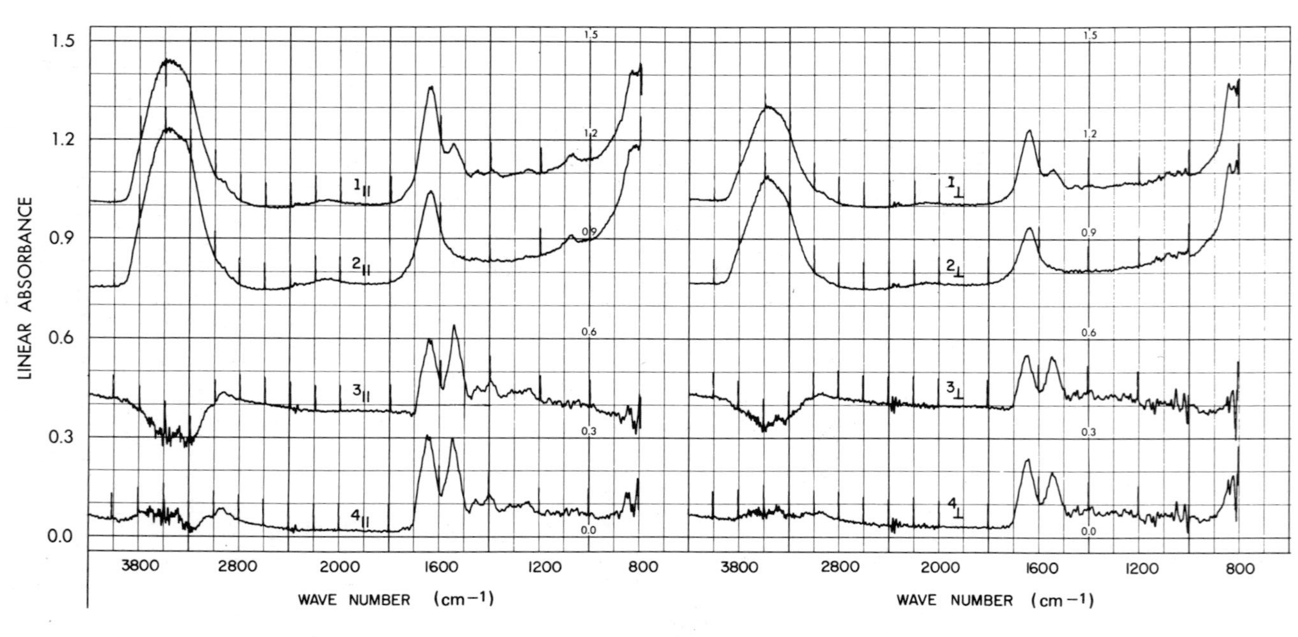

FIG. 18. Spectra $1_{\parallel}$ and $1_{\perp}$: 0.375X ordinate-expanded, once smoothed average of four scans of 0.36% solution of 72% clottable bovine fibrinogen in pH 7.2, 0.10 M phosphate buffer. Spectra $2_{\parallel}$ and $2_{\perp}$: background spectra of buffer solution only, as in $1_{\parallel}$ and $1_{\perp}$. Spectra $3_{\parallel}$ and $3_{\perp}$: direct subtraction of spectra $2_{\parallel}$ and $2_{\perp}$ from $1_{\parallel}$ and $1_{\perp}$, 1.0X ordinate expansion, twice-smoothed. Spectra $4_{\parallel}$ and $4_{\perp}$: scaled subtraction of $0.90 \cdot 2$ and $0.88 \cdot 2$ from $1_{\parallel}$ and $1_{\perp}$, respectively, 1.0X ordinate expansion, twice smoothed.

at 3400 cm^{-1} in spectra $3_{\parallel}$ and $3_{\perp}$ is due to the fact that protein molecules have replaced water in the region near the germanium surface. To correct this, spectra $4_{\parallel}$ and $4_{\perp}$ are scaled absorbance subtractions, removing 90% and 88% of the buffer baselines, respectively. The final results are the protein spectra $4_{\parallel}$ and $4_{\perp}$, clearly showing the amide I band at 1640 cm^{-1} which is normally obscured by the 1639 cm^{-1} water band ($1_{\parallel}$ and $1_{\perp}$).

C. Using Recorded Spectra in FORTRAN or BASIC Programs

By formatting the absorbance data as floating point numbers, four bytes per datum, spectra recorded using the RSMAS program are compatible with Data General's FORTRAN IV and BASIC languages. The ASCII title line associated with each cassette file can be read as a string literal in BASIC provided that the title line is enclosed in quotation marks.

The FORTRAN and BASIC compatibility of the recorded spectra has proven extremely useful in applications involving peak separation and integration, and in the use of "canned" FORTRAN routines available from the literature or from larger computer centers. For instance, all of the routines described in NRC Bulletins 11, 12, and 13 [2-4] are thus readily adapted to the NOVA for use in handling spectra.

ACKNOWLEDGMENTS

We wish to acknowledge the National Institutes of Health, NHLI Grant HL-15919-01A1, for the financial support necessary to develop this system, as well as the technical assistance and advice of Dr. Robert Hannah of Perkin-Elmer Corporation and Ken Brandt of Data General Corporation.

REFERENCES

1. H. A. Laitinen, The seven ages of an analytical method, *Anal. Chem.*, *45*:2305, 1973.
2. R. N. Jones, T. E. Bach, H. Fuhrer, V. B. Kartha, J. Pitha, K. S. Seshadri, R. Venkataraghavan, and R. P. Young, Computer

programs for absorption spectrophotometry, Bulletin No. 11, National Research Council of Canada, Ottawa, 1968.

3. J. Pitha and R. N. Jones, Optimization methods for fitting curves to infrared band envelopes: Computer programs, Bulletin No. 12, National Research Council of Canada, Ottawa, 1968.

4. R. N. Jones and R. P. Young, Additional computer programs for absorption spectrophotometry and band fitting, Bulletin No. 13, National Research Council of Canada, Ottawa, 1969.

5. J. S. Mattson and A. C. McBride III, Digital data acquisition from a Hilger-Watts H-1200 infrared spectrophotometer, *Anal. Chem.*, *43*:1139 (1971).

6. R. W. Hannah and A. Savitzky, Rules for digital smoothing of spectral data--Part I, presented at the Eastern Analytical Symposium, Atlantic City, November, 1972.

7. A. Savitzky and R. W. Hannah, Rules for digital smoothing of spectral data--Part II, presented at Pittsburgh Conference on Anal. Chem. and Appl. Spectry., Cleveland, March, 1973.

8. A. Savitzky and M. J. E. Golay, Smoothing and differentiation of data by simplified least squares procedures, *Anal. Chem.* *36*:1627 (1964).

9. J. Steiner, Y. Termonia, and J. Deltour, Comments on smoothing and differentiation of data by simplified least square procedure, *Anal. Chem.*, *44*:1906 (1972).

10. R. G. J. Miller and B. C. Stace (eds.), *Laboratory Methods in Infrared Spectroscopy*, 2nd ed., Heyden, London, 1972.

11. H. Hansdorff and H. Sternglanz, Differential recording with infrared spectrophotometers, *Suppl. Vol. I, Ser. X, Del Nuovo Cimento*, *1955*:1, Perkin-Elmer Reprint No. IR-13.

12. F. S. Parker, *Infrared Spectroscopy in Biochemistry, Biology, and Medicine*, Plenum, New York, 1971.

13. N. J. Harrick, *Internal Reflection Spectroscopy*, Wiley-Interscience, New York, 1967.

Chapter 3

TRACE GAS DETECTION BY CORRELATION SPECTROSCOPY

R. H. Wiens

H. H. Zwick*

Barringer Research, Ltd.
Rexdale, Ontario, Canada

*Current affiliation: Canada Centre for Remote Sensing, Ottawa, Ontario, Canada

I. INTRODUCTION

A. Descriptive

One of the more recent techniques of analytical chemistry is correlation spectroscopy. The spectrum of incoming light is correlated with the spectrum of a specific target gas or gases which is in some way stored in the correlation spectrometer. The spectrum can be seen in absorption or emission, and sensors described in this chapter have been built for either or both. Many of the recent advances in correlation spectroscopy have been made at Barringer Research Limited (BRL) and it is this work which is discussed here. The closely related technique of Hadamard transform spectroscopy is not included, nor is derivative spectroscopy since both of these are described adequately by others (e.g., Hager, 1973; Harwit and Decker, 1975).

Three types of correlation spectrometers are treated. In its simplest form the dispersive type is a grating spectrometer in which the exit slit is replaced by a series of slits which corresponds to the spectral absorption features of the target gas spectrum. A second mask covers the same spectral region but has slits placed to view the nearby background spectrum with minimum influence of the target spectrum. These two correlator masks are alternately placed in the exit plane of the spectrometer so that the detector receives a modulated signal. The depth of the modulation is a measure of the strength of the target gas spectrum in the incoming light. The detector output is electronically demodulated to produce a dc voltage proportional to the amount of gas within the field of view of the instrument. In the case of periodic spectra, the second mask may be identical to the first but displaced by a half-period. A dispersive correlation spectrometer was first used by Bottema et al. (1964) to measure the water vapor content of the atmosphere of Venus. It was later applied to measurements of SO_2 in the earth's atmosphere by Barringer and Schock (1966) and Kay (1967).

The second correlation spectrometer is the gas filter type, in which a filter having a transmission function identical to the spectrum of the target gas serves as the correlating element. Such a filter is readily available in a sample of the target gas itself. Incoming light passes through such a sample gas filter cell and a similar reference cell containing either no gas or gas with no spectral features in the region of the target gas spectrum. When these two cells are alternately inserted between the source and the detector, the detector produces a modulated signal, the depth of which again depends upon the amount of target gas in the sensor field of view. Nondispersive analyzers based on this principle have long been available commercially, starting with patents by Schmick in 1926 and Luft in 1938 (Vasco, 1968). More recent versions have been reported by Goody (1968), Ludwig, et al. (1969), and Ward and Zwick (1975).

The correlation interferometer, the third type of correlation spectrometer, is a Michelson interferometer, incorporating its advantages, but using a predetermined interferogram stored in the instrument for correlation with the current interferogram. This eliminates the need for transforming the interferogram back to the spectrum and permits more rapid data interpretation than the conventional Fourier spectrometer. Only that portion of the interferogram with strong target gas structure is scanned, reducing the need for large path difference scanning range and increasing detector sensitivity. The amplitude of the interferogram depends upon the amount of target gas in the light path. A sophisticated signal processing technique, integrating the product of the interferogram envelopes and the correlating function (itself a specially chosen interferogram) is employed to eliminate the effects of gases with competing spectra. The first report of a correlation interferometer of this type was by Dick and Levy (1970).

One might ask in what sense the correlation spectrometers represent an advance over earlier instruments for spectrochemical analysis, especially the scanning spectrometer with a standard light source. It is helpful to answer in terms of the advantages of a Michelson interferometer over the scanning spectrometer. They are the Jacquinot advantage, that the interferometer has greater throughput or "etendue" than any spectrometer of similar resolution; and the Felgett advantage, that all spectral features are observed simultaneously. The Jacquinot advantage is automatically a feature of the correlation interferometer and the gas cell correlation spectrometer. The throughput of the grating correlation spectrometer is increased by the use of multiple entrance slits. These n slits produce n identical spectra displaced from each other in such a way that the transmission maxima coincide. The throughput is better than that of a single slit spectrometer by a factor of n, but of course this advantage is only attainable with periodic spectra. The multiple exit slits ensure that all the significant spectral features are viewed simultaneously, giving the correlation spectrometer a significant multiplex (or Felgett) advantage over the scanning spectrometer as well. The multiplex advantage of the gas cell correlation spectrometer is even better than that of the dispersive type (because the spectral matching is more detailed), and that of the correlation interferometer is the same as that of any Michelson interferometer. One of the main benefits of correlation spectroscopy is the simplicity of the data which are produced. The versatility of the scanning spectrometer and the Fourier spectrometer are sacrificed in order to make the instrument output proportional to the concentration of the target gas. By dedicating the correlation spectrometer to the measurement of only one or two gases at a time it has been possible to combine high sensitivity with on-line data reduction to make a family of sensors ideal for field use.

The purpose of the BRL family of correlation spectrometers has been the detection of atmospheric pollution primarily (but not necessarily) using natural light sources. Dispersive correlation spectrometers have been built and used for the detection of SO_2, NO_2, and I_2, all of which have relatively periodic spectra in the visible and near-ultraviolet regions of the spectrum. Gas-filter correlation spectrometers have been used in the infrared for CO, CH_4, C_2H_6, HCl, CO_2, and SO_2, and work is in progress on other gases as well. The correlation interferometer has been used primarily for CO and

CH_4 detection, but NH_3, NO_2, N_2O, CO_2, and H_2O are also feasible atmospheric target gases.

Atmospheric pollution can be measured actively using a controlled artificial light source, or passively with whatever natural light sources are available. Active sensing is in general much simpler than passive, even if less convenient, because the spectral characteristics of the source are known. A number of active sensors have been built, and they will be referred to later. Passive sensing is difficult because the natural source imposes its own spectrum upon that of the target gas. The source spectrum is usually variable, and, since the source itself is inaccessible, it cannot be measured independent of the intervening atmosphere. Consideration of the possible light sources and all their anticipated variations is therefore one of the essentials of the instrument design.

B. The Bouguer-Beer Absorption Law

The basis of all quantitative measurements in absorption spectroscopy is the Bouguer-Beer law of absorption.

$$dN(\lambda) = -N(\lambda)k(\lambda)p(x)\ dx$$

When light of radiance $N(\lambda)$ at wavelength λ enters an absorbing medium of absorptivity $k(\lambda)$ at partial pressure $p(x)$, it decreases by an amount $dN(\lambda)$ after traversing a thickness dx. When integrated over a path length L this equation gives the radiance $N(\lambda)$ at L in terms of the source radiance $N_o(\lambda)$

$$N(\lambda) = N_o(\lambda)\ \exp\left(-k(\lambda)\int_0^L p(x)\ dx\right) \tag{1}$$

The integral $\int_0^L p(x)\ dx$ is frequently given in units of atmosphere-centimeter when $k(\lambda)$ is given in $cm^{-1}\ atm^{-1}$. For sensing pollutant gases in the lower troposphere the partial pressure is on the order of microatmospheres and is equivalent to a volume concentration $c(x)$ in parts per million (ppm).

When measurements are made in the active mode the path length L is known. If the medium is homogeneous, the integral reduces to cL and c can be found from the equation by measuring $N(\lambda)$ and $N_o(\lambda)$.

If the variation of c with x is not known, an average concentration over the distance L may be obtained. In passive sensing both the length L and the distribution of pollutant c(x) are unknown and only the whole integral can be found from Eq. (1). The integral is nevertheless written cL for convenience and given the name burden with the unit of parts per million-meter (ppm-m). A passive sensor measures burden only and to reduce this to a concentration requires an assumption about the distribution of target gas along the absorption path.

When a number of absorbing gases are present the law of additivity applies, i.e., the optical thickness of the mixture is the sum of the optical thicknesses of the individual gases.

$$N(\lambda) = N_o(\lambda)\ \exp\left(-\Sigma k_i(\lambda) \int c_i(x)\ dx\right)$$

The resultant spectrum is thus the product of the spectra of the constituent absorbers.

It is worthwhile to note that the Bouguer-Beer law is strictly true only for monochromatic radiation. In expressing the response of an instrument in terms of the gas burden it is sometimes mathematically convenient to employ average absorptivity over a range of wavelengths. The total intensity does not obey the simple exponential law because

$$\int N_o(\lambda)\ \exp\left[-k(\lambda)cL\right]\ d\lambda \neq \exp\left[-cL\ \frac{\int k(\lambda)\ d\lambda}{\int d\lambda}\right] \int N_o(\lambda)\ d\lambda$$

The approximation is worse the greater the spread of k(λ). Average absorptivities are therefore used only in predicting the form of the response curves, which are in practice always determined empirically.

The major difficulty in using the Bouguer-Beer law for passive sensing in the atmosphere is that the source spectrum $N_o(\lambda)$ is not generally known. It can not be measured once for all because it is continuously changing, and it cannot be correctly measured remotely (by the correlation spectrometer) without making an assumption about the absorbers between source and sensor. The problem has not been solved, but much effort has been invested in minimizing its effect. For example, the dispersive correlation spectrometer looks at two sets of wavelengths, λ_1 and λ_2. Then

$$\frac{N(\lambda_1)}{N(\lambda_2)} = \frac{N_o(\lambda_1)}{N_o(\lambda_2)} \exp[-k(\lambda_1) - k(\lambda_2)cL]$$

This reduces the need to know the absolute source spectrum at all times to knowing the relative intensities or the shape of the spectrum. Spectral shape may be expected to be less variable than absolute intensity and once measured might be applicable at future times. Unfortunately the shape of the source spectrum does vary particularly in the visible and ultraviolet regions of the spectrum and other devices must also be employed.

C. Passive Atmospheric Sensing

1. Sources of Natural Light

The radiation sources that can be used in detecting trace gases in the atmosphere are the sun, the sky, and the earth. The sun is the most intense source, and it emits a wide range of wavelengths, but from the point of view of a ground-based observer it is also inconvenient. Radiation from the earth's surface is considerably less uniform, but for observations from an airplane or satellite it is often especially useful. It consists of a reflected and an emitted component. The reflected component comes initially from direct sunshine or from diffuse skylight; its spectrum depends, therefore, upon the spectra of the sun and sky as well as the spectral reflectivity of the surface. The emitted component is of thermal origin and has the spectral characteristics of a blackbody at the temperature of the surface modified by the spectral emissivity of the surface. Skylight, too, has a thermal emission component described by the blackbody radiation characteristics of the atmospheric temperature and the emissivity of the constituent gases and aerosols. It also has a scattered component, which originates from direct sunlight, reflected light from the ground, and light from other parts of the sky, all of which are scattered from molecular air and from aerosols. The relative intensity of any component depends upon the wavelength region which is of interest. The sunlight reflected from the ground is equal to the light emitted by the earth and the atmosphere at a wavelength of approximately 3 μm depending of course upon local conditions. At wavelengths greater than 4 μm the reflected component is usually negligible; at wavelengths less than 2 μm the thermal emission component is negligible.

In all spectral regions the Bouguer-Beer law is merely an approximation to the radiative transfer equation, discussed for the earth's troposphere by Goody (1964). When light of radiance $N(\lambda)$ traverses an atmospheric element of optical thickness $d\tau$, where

$$d\tau = k(\lambda)p(x)\ dx$$

the change of radiance is given by

$$dN(\lambda) = -N(\lambda)\ d\tau + J(\lambda)\ d\tau \tag{2}$$

The first term on the right accounts for absorption and scattering out of the beam. The second term is the amount added to the beam by emission from or scattering by the element $d\tau$. It is called the source function, and when it is ignored the radiative transfer equation reduces to the Bouguer-Beer law. Integration of Eq. (2) to evaluate cL in terms of $N(\lambda)$ requires some knowledge of the source function. Unfortunately a generalized source function including the effects of emission and scattering at all wavelengths is complicated. The natural division between thermal and nonthermal spectral regions is useful here.

The intense radiation visible in all parts of the sky attests to the importance of scattering in the uv-visible region. Goody (1964) shows that the source function is proportional to the integral of the product of the Poynting vector of the incident radiation and the phase matrix of the scattering atmosphere. The phase matrix can be computed from electromagnetic theory using the Rayleigh and the Mie approaches. A good discussion of theoretical and experimental aspects of the scattering function is given by Bullrich (1964) and numerous computations of the daylight spectrum have been published (e.g., Braslau and Dave, 1973). However, each separate use of the sensor requires an individual solution of Eq. (2), which renders these methods too cumbersome for routine data reduction. Fortunately the scattering term is small enough to ignore in many pollution sensing situations, and the Bouguer-Beer law applies. In specific cases methods can be devised to compensate. For example, Moffat and Millan (1971) presented a method to correct observations on distant smoke-stack emissions for dilution by scattering.

The source function in the infrared may be written

$$J(\lambda) = \varepsilon(\lambda)B(\lambda,T)$$

where $\varepsilon(\lambda)$ is the atmospheric spectral emissivity and $B(\lambda,T)$ is the Planck radiance function at temperature T. Even this simple source function presents difficulties in solving Eq. (2) unless some simplifying assumptions are made, as shown in the following subsection.

2. Infrared

Most pollutants of interest have fundamental rotation-vibration band spectra in the wavelength region beyond 2 μm, and, as already noted, the observations of these pollutants require that the Bouguer-Beer law be augmented by the source function $\varepsilon(\lambda)B(\lambda,T)$; the full equation of radiative transfer is necessary. The emissivity is defined as the ratio of radiance emitted from a body, $N(\lambda)$, to that emitted by a blackbody at the same temperature, $B(\lambda,T)$. According to Kirchhoff's law the radiance emitted by a body in equilibrium with its surroundings is equal to the radiance absorbed by the body. Provided that one can neglect induced emission, the fraction of the radiance transmitted by an optical thickness τ is $\exp(-\tau) \equiv t$, which is called (Armstrong and Nicholls, 1972) the transmission function. Thus the fraction absorbed is $1 - t$, and this is equal to the emissivity. Equation (2) may now be written

$$dN(\lambda) = -N(\lambda)\ d\tau + \varepsilon(\lambda)B(\lambda,T)\ d\tau$$

This differential equation is linear in $N(\lambda)$ and could be solved by the use of an integrating factor $\exp(\tau)$. In order to integrate the right-hand side of the equation some knowledge of $B(\lambda,T)$ as a function of optical path τ is required. This implies that the temperature profile of the atmosphere along the line of sight be known. The simplest case is that of an isothermal atmosphere where $B(\lambda,T)$ is constant and can be removed from the integrand. This situation is an oversimplification and is not expected to exist in a real sense. However, some helpful insights into infrared remote sensing can be gained from the result.

The final expression, for the radiance reaching the sensor in this isothermal atmosphere when scattering and induced emission are insignificant, is

$$N(\lambda) = N_o(\lambda)t + (1 - t)B(\lambda,T)$$

where $N_o(\lambda)$ is the background radiance. Often it is useful to express the background radiance in terms of its blackbody radiation characteristics $N_o(\lambda) = \varepsilon_b(\lambda)B(\lambda,T_b)$

$$N(\lambda) = t_g(\lambda)\varepsilon_b(\lambda)B(\lambda,T_b) + \varepsilon_g(\lambda)B(\lambda,T_g) \quad (3)$$

where b and g indicate background and target gas, respectively.

Equation (3) shows that if the radiance of the background exceeds that of the target gas, the gas can be detected in absorption. If the reverse is true it is seen in emission. For measurements against a clear sky the background radiance is negligible and only the emission term applies (supposing the diffusely scattered sunlight to be negligible, as is true, beyond 3 μm). This provides the fortunate circumstance that a passive infrared sensor is as useful at night as in the day. If the measurement is made using the distant hot sun as background source, the absorption term is much larger than the atmospheric emission term. This also describes the situation using an active source. If the sky is overcast, measurement is still possible because the radiance of the cloud cover is rarely the same as the radiance of the intervening atmosphere. It is clear, however, that now an absorption and emitting process are competing with the result that sensitivity to the presence of target gas is reduced.

Observations from an aircraft or space platform toward the earth are slightly more complicated than sky measurements. At wavelengths greater than 4 μm the sunlight reflected back from the ground is negligible, and whether the gas is detected in absorption or emission depends upon whether the ground radiance is larger or smaller than the gas radiance. Typically the ground is warmer than the air during the day and cooler at night causing absorption and emission to alternate. This means that twice each day it is possible for $\varepsilon_b(\lambda)B(\lambda,T_b) = B(\lambda,T_g)$. When this happens Eq. (3) becomes

$$N(\lambda) = B(\lambda,T_g)$$

Then the radiation arriving at the sensor does not contain the spectrum of the target gas and measurement is impossible. The mathematical independence of $N(\lambda)$ from τ finds physical explanation in the gas emitting as many photons as it absorbs. This condition does not occur when the temperature of the gas and background are equal unless the background is perfectly black ($\varepsilon_b = 1$). The dependence upon emissivity has been suggested by Zwick and Millan (1971) to make observation possible at virtually all times, because small scale or local ground emissivity may generally be expected to change quickly with position over most land areas.

The spectral radiance for wavelengths in the 2-4 μm range for both the sky and the earth are shown in Figs. 1 (Bell et al., 1960) and 2 (Wolfe, 1965), respectively. The radiance at wavelengths less than 3 μm is due to sunlight scattered in the sky or reflected from the ground. Observations in this range are in absorption during the day.

3. Visible-Ultraviolet

In the visible and ultraviolet spectral region the target gas must always be observed in absorption because the energies of the molecular transitions involved are far greater than can be attained thermally at terrestial temperatures. The solar spectrum is thus of primary concern. It is shown in Fig. 3 at low resolution both above and below the atmosphere. The extraterrestrial spectrum is well

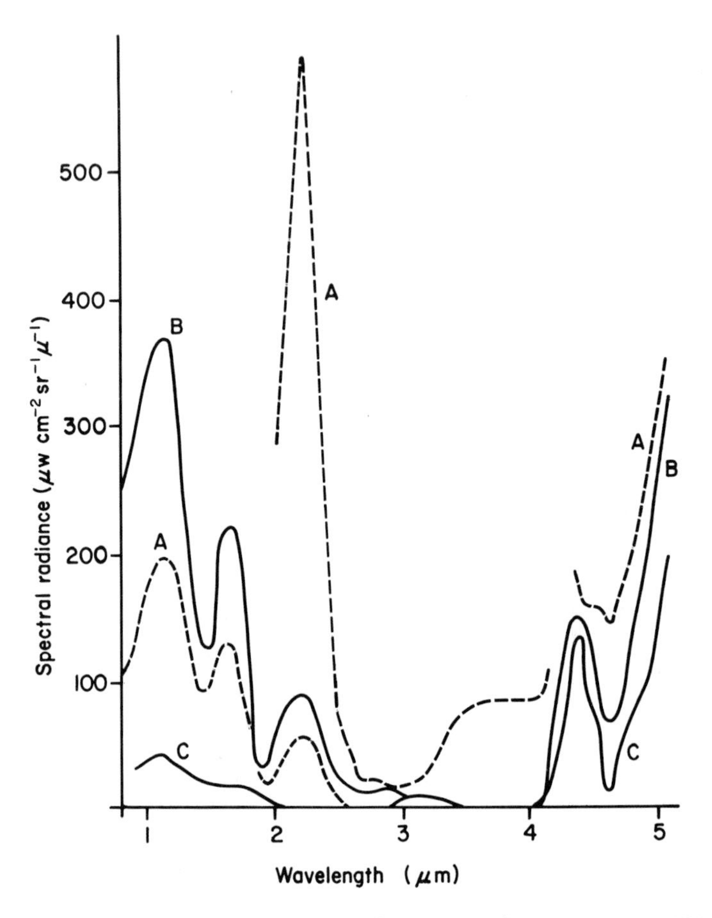

FIG. 1. Spectral radiance for sky. (Bell et al., 1960)

approximated by the 5900 K blackbody radiation curve. The sea level solar spectrum has superposed on it wide absorption regions caused primarily by ozone, oxygen, carbon dioxide, and water vapor. In order for solar radiation to be used for pollution measurements the target gas must have absorption bands within the atmospheric windows, i.e., between the major absorption regions of Fig. 3. All of these windows are somewhat variable because of the varying geometrical path of the sunlight through the absorbing layers and because of the varying amounts of absorber in the atmosphere.

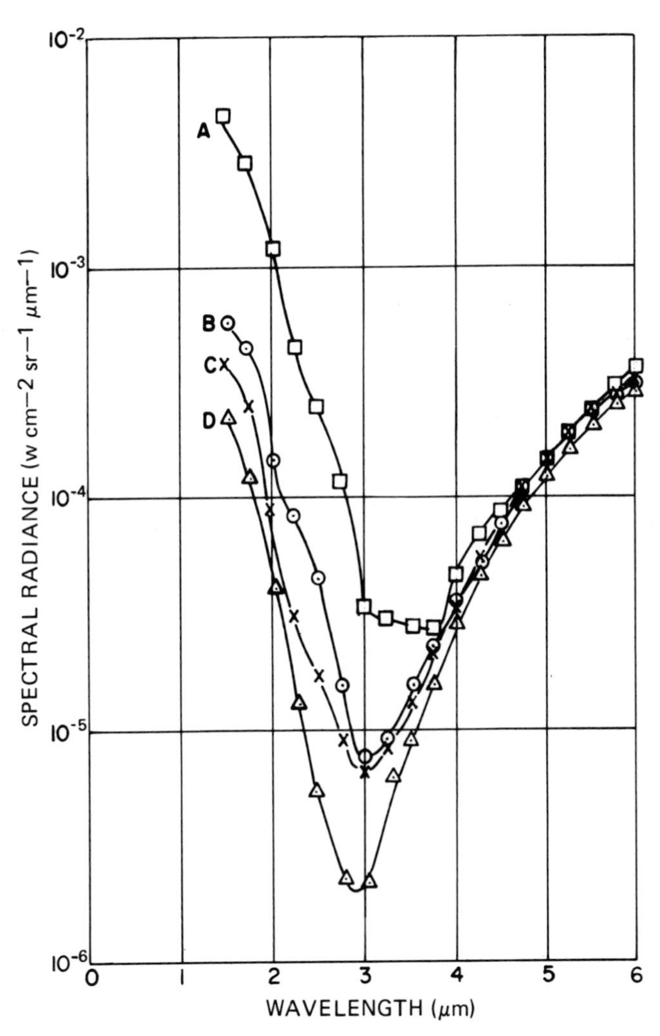

FIG. 2. Spectral radiance for earth. (Wolfe, 1965)

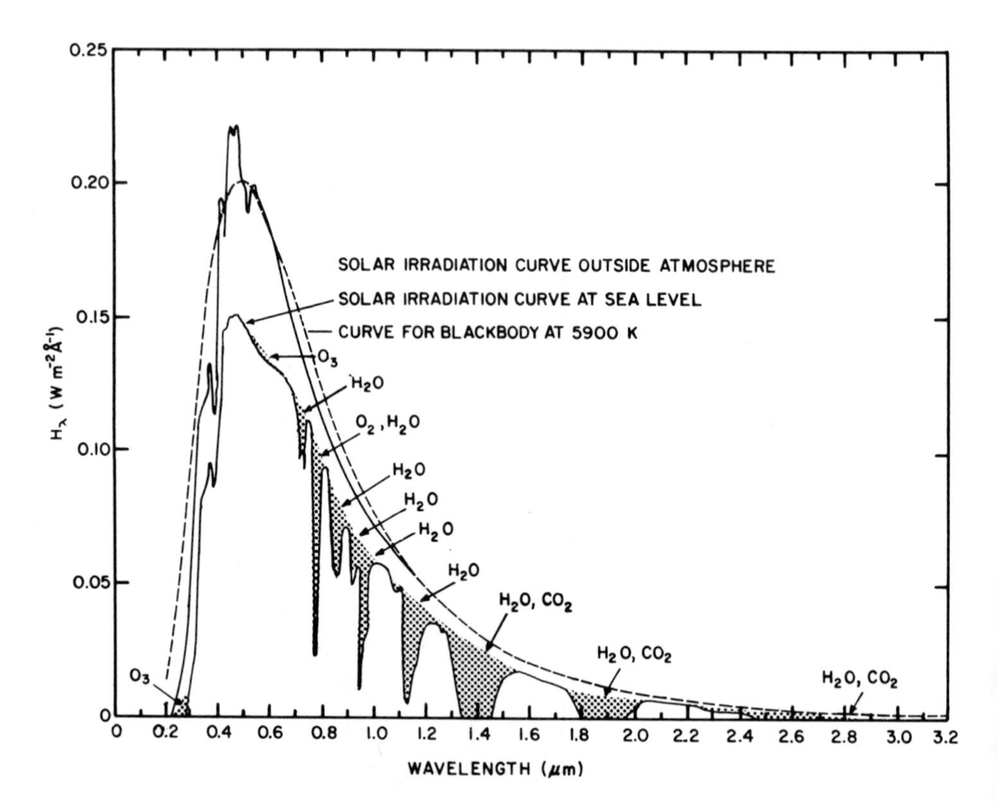

FIG. 3. Solar spectrum at low resolution. (Kondratyev, 1969)

A particular case is the ultraviolet limit of observation, of special interest because it is in the vicinity of the SO_2 absorption bands. The ultraviolet limit is set by the optical thickness of the ozone layer through which the sunlight must pass and is of course dependent upon the sensitivity of the instrument used for detection. Table 1 summarizes the results of Poliakova (Kondratyev, 1969) of the USSR, who measured the wavelength limit of detection with a particular spectrometer as a function of solar elevation angle. The table implies regular diurnal and annual variations but these are further complicated by the less regular diurnal and annual variations of the total ozone content in the vicinity of the observer.

The variations that occur in the spectrum of direct solar radiation as seen from below the atmosphere are imposed upon the more

TABLE 1

Ultraviolet Limit of Solar Spectrum

Solar altitude (deg)	Minimum wavelength (nm)
1	420
2	382
3	352
5	327
7	318
10	312
15	306
20	304
25	302
30	300
35	298
40	297
45	296
50	295

useful diffuse light, but additional spectral changes also occur. The complex nature of the daylight spectrum is most simply seen in the varying shades of blue in the clear sky and the varying shades of gray in the overcast sky. The scattering cross section of air is given by the Rayleigh law and is proportional to λ^{-4}. Aerosol scattering is computed on the basis of Mie theory and depends upon the size distribution of the particles. Bullrich (1964) found that when the visibility is 1 km or greater, the scattering cross section for aerosols is proportional to λ^{-1}. For very low visibility, such as in a fog or cloud, the cross section becomes independent of λ. As already indicated, the scattering laws depend not only upon wavelength, but also upon the direction of the incident light, that is, the position of the sun. The highly variable aerosol number and size distribution coupled with the continuously changing solar angle add a similarly variable scattered component to the more regularly

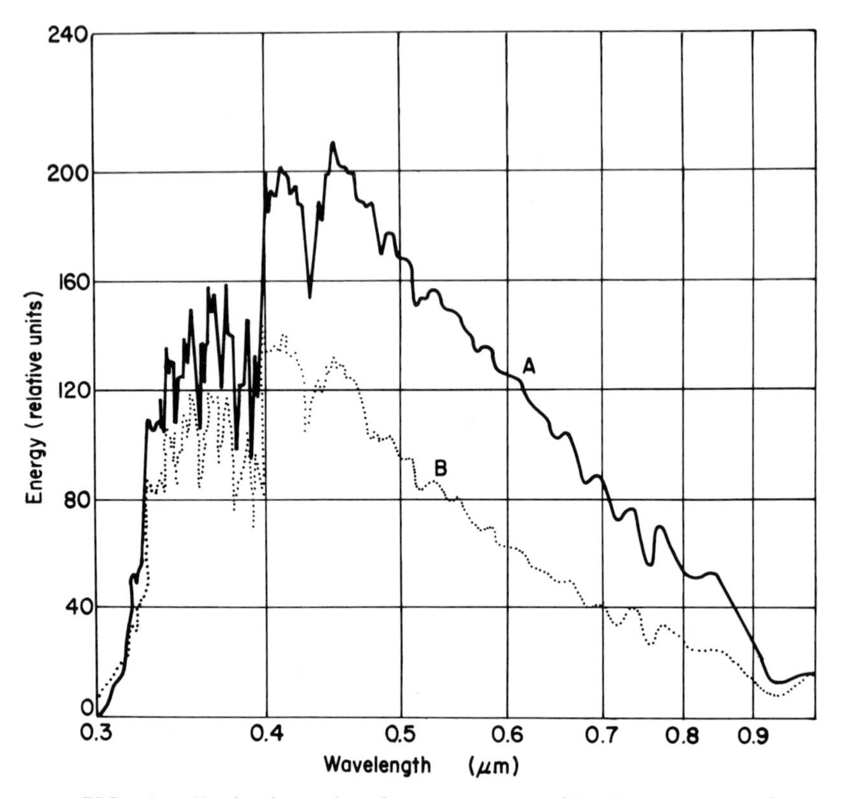

FIG. 4. Variations in sky spectrum. (Kondratyev, 1969)

varying Rayleigh component. Also, some of the skylight comes from the ground reflecting back the skylight with the further imposition of its own reflection spectrum and some solar stimulated surface emission. An example of the variability of the sky spectrum is shown in Fig. 4 by Lenz (Kondratyev, 1969). The graph permits comparison of the spectrum at the zenith and at the point of minimum intensity for the same clear sky. Note that the two traces cross at 320 nm.

A further complicating feature of the sky spectrum is the abundance of Fraunhofer lines and bands in the visible region. That the Fraunhofer lines are subject to a filling-in effect when viewed in skylight relative to those seen in the direction of the sun was demonstrated by Noxon and Goody (1965). Harrison and Kendall (1974) showed that the percentage of in-filling of the CaI line at 4227A decreased with increasing solar zenith angle for perfectly clear skies, but that it varies unpredictably when haze or cloud appear anywhere in the sky. The suggested explanations are rotational Raman scattering by atmospheric gases, aerosol fluorescence, and damped molecular Rayleigh scattering from the ground, none of which is strongly dependent upon wavelength. Thus the effect of the in-filling is the addition of a variable but continuous spectral component to the daylight spectrum.

It is clear that any remote sensor using the sky as a spectral source in the uv-visible range must be able to distinguish between target gas variations and background variations. So far only the dispersive correlation spectrometer has been designed for uv-visible observations. The methods used to overcome the problems of background are discussed in the following sections where each instrument is discussed in more detail.

II. DISPERSIVE CORRELATION SPECTROSCOPY

A. General

Although all correlation spectrometers have certain common characteristics, the differences among them are great enough to warrant separate treatment for each. The dispersive type is considered first because it is most directly related to more traditional spectrometers. It is used to best advantage with spectra having periodic band structure such as that of I_2 shown in Fig. 5. The technique requires first that the incoming light be dispersed by a grating into its spectrum and then that the spectrum be transmitted through a mask placed in front of the detector. This mask is made to shift from the transmission wavelength maxima of the spectrum to an adjacent wavelength set; that is, the mask takes the two positions shown in Fig. 5. The difference between the detector responses for the two mask positions divided by the response to the maximum position gives the average fractional depth of the absorption spectrum or the average transmittance of the target gas material.

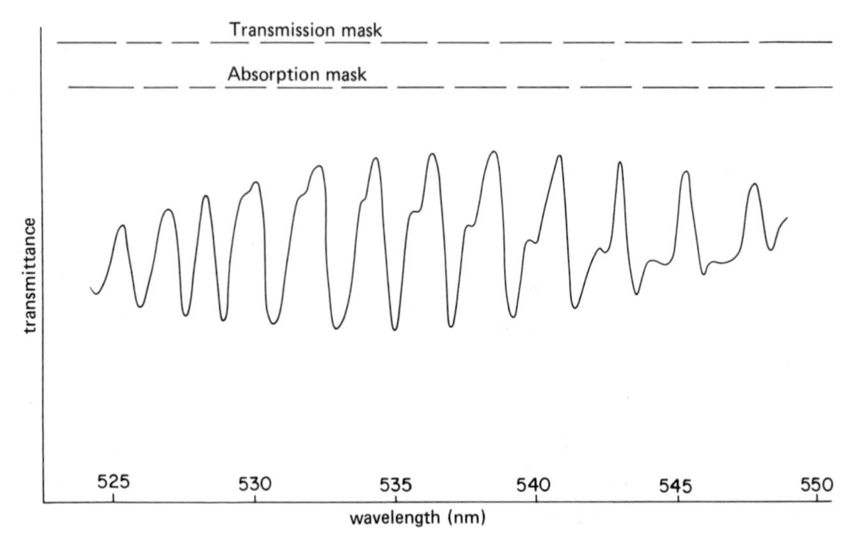

FIG. 5. Dispersive correlation spectrometer with periodic band structure.

In this section we will first derive the response of a dispersive correlation spectrometer to the presence of target gas, then calculate the signal-to-noise ratio that such a device can provide. The mask optimization procedure is then described in terms of the signal-to-noise ratio, and the embodiment of the principles is demonstrated by examples, primarily the correlation spectrometer presently in production at Barringer Research. The mathematical analysis and mask optimization procedures described here are those given by Millan (1972).

B. Signal Processing

1. Instrument Response

When light falls on a photomultiplier through an optical sensor the power available at one wavelength is given by

$$p(\lambda) = A\Omega\beta(\lambda)N'(\lambda)$$

where

$p(\lambda)$ is spectral radiant power ($W \cdot nm^{-1}$)

$A\Omega$ is system throughput ($cm^2 \cdot sr$)

$\beta(\lambda)$ is fraction of light transmitted by system or optical transfer function

$N'(\lambda)$ is intensity of light reaching instrument $(W \cdot m^{-2} \cdot sr^{-1} \cdot nm^{-1})$

Substituting the Bouguer-Beer law from Sec. I we have

$$p(\lambda) = A\Omega\beta(\lambda)N(\lambda)\exp[-k(\lambda)cL]$$

using $N(\lambda)$ to represent the source spectrum and any spectral interference.

The total power falling on the detector is the integral of the convolution function between the spectral radiant power $p(\lambda)$ and the mask function $F(\lambda)$ over the mask position ξ. The mask function in turn is the convolution of the entrance slit function of the spectrometer with the exit slit function and is generally a series of triangles. (A specific example is given in Sec. II.C.) Thus the radiant power on the detector with the mask in position ξ is

$$P(\xi) = \iint_{-\infty}^{\infty} p(\lambda)F(\lambda - \xi)\, d\lambda\, d\xi$$

Since at least two positions of the same mask or at least two masks are necessary for the technique, we write for each of j masks

$$P_j = \iint_{-\infty}^{\infty} p(\lambda)F_j(\lambda - \xi)\, d\lambda\, d\xi$$

$$= A_j\Omega \iint_{-\infty}^{\infty} N(\lambda)\beta(\lambda)F_j(\lambda - \xi)\exp[-k(\lambda)cL]\, d\lambda\, d\xi \qquad (4)$$

The photomultiplier converts this power to a cathode current I_k, which for the jth mask is

$$I_{kj} = P_j S_k + I_{k0}$$

I_{k0} is the cathode dark current, and S_k is the detector responsivity in units of A/W. The wavelength dependence of S_k can be included in the optical transfer function $\beta(\lambda)$. The anode current is dependent upon the potential difference across the photomultiplier. In the Barringer series of instruments this potential difference is controlled by an automatic gain control (AGC) which maintains the anode current of the first mask at some preset value I_1, and holds this gain constant during the second and subsequent mask measurements, until the first mask value is remeasured. This introduces a variable gain G which applies to each of the j mask positions. The anode currents are

$$I_j = GI_{kj} = (P_j S_k + I_{k0})G = P_j S_k G + I_0 \tag{5}$$

The last term implies that all the anode dark current originates as cathode dark current with no contribution from the dynode chain, but this limitation is not serious. We can now use this equation to compute the gain, for it also applies to the first mask.

$$G = \frac{I_1 - I_0}{P_1 S_k}$$

The signal processing electronics forms the difference between the two anode currents

$$\Delta I = I_1 - I_2 = (P_1 - P_2) S_k\ G = \left(1 - \frac{P_2}{P_1}\right) P_1 S_k G$$

or

$$\Delta I = \left(1 - \frac{P_2}{P_1}\right)(I_1 - I_0)$$

This difference is detected through a resistance r and integrated over one cycle of operation. The pulse for each mask is present only for a fraction d of the total duty cycle so that the integral is only d times as large as each pulse. Finally, we assume that enough light is available to hold I_1 to a value such that $I_1 >> I_0$. The resulting voltage response is therefore

$$V = rd\left(1 - \frac{P_2}{P_1}\right) I_1 \tag{6}$$

The complete response equation is found by substituting the values of P_j from Eq. (4) into Eq. (6). Some simplification is possible by designing the mask set areas A_j so that the throughput, $A\Omega$, is the same for both mask positions and hence cancels in Eq. (6). Next we introduce average absorption coefficients a_j which enables us to remove the exponentials from the integrands. Finally, we introduce the ratio

$$\psi = \frac{\iint_{\infty}^{\infty} N(\lambda)\beta(\lambda)F_2(\lambda - \xi)\, d\lambda\, d\xi}{\iint_{-\infty}^{\infty} N(\lambda)\beta(\lambda)F_1(\lambda - \xi)\, d\lambda\, d\xi}$$

With all these simplifications Eq. (6) becomes

$$V = I_1 rd\{1 - \psi \exp[(a_1 - a_2)cL]\}$$

which is equivalent to

$$V = I_1 rd(1 - \psi) + I_1 rd\ \psi\{1 - \exp[a_1 - a_2]cL\} \qquad (7)$$

The first term of Eq. (7) is the response of a two-mask instrument if $cL = 0$, and is called the zero-gas term. The equation thus states that the response is the sum of a gas-dependent and a zero-gas term, both of which contain ψ. For any type of constant source ψ is constant and the zero gas term implies a constant offset which can be removed electronically. The ψ in the second term is merely a scaling factor which can be fully accounted for by calibration. Thus, for stable spectral sources a two-mask correlation spectrometer offers unambiguous response to fixed amounts of target gas. When used in the passive mode with skylight as the spectral source, however, Eq. (7) does not provide an unambiguous response. In this situation ψ changes with time, adding a generally unknown amount to the zero-gas term and shifting the baseline. These baseline changes caused by sky spectrum variations are indistinguishable from changes added to the second term by the presence of gas. Because of this difficulty with Eq. (7) more advanced correlation spectrometers used for passive remote sensing employ four masks. The first pair yield a voltage response given by Eq. (7). To calculate the response of the second pair we proceed with a similar analysis. The pulse of anode current caused by the third mask is held by a second AGC to the same magnitude as that of the first mask.

$$I_3 = P_3 S_k G_2 + I_0 = I_1$$

and

$$\Delta I = I_3 - I_4 = \left(1 - \frac{P_4}{P_3}\right) S_k G_2 P_3 = I_1 \left(1 - \frac{P_4}{P_3}\right)$$

Again the difference between the two anode currents corresponding to the two mask positions of the second pair (3 and 4) is taken electronically and continued with similar electronic processes. Repeating the same analysis for this pair that led to Eq. (7) gives

$$V' = rI_1 d(1 - \psi') + rI_1 d\ \psi\{1 - \exp[(a_3 - a_4)cL]\} \tag{8}$$

The difference between the responses of the two pairs of masks is now taken

$$R = V - V'$$

$$R = rdI_1\psi\{\exp[(a_3 - a_4)cL] - \exp[(a_1 - a_2)cL]\} \tag{9}$$

The advantage of Eq. (9) over (8) is that the zero gas term is no longer present and consequently ψ dependent baseline drifts are eliminated. Of course this is only true if $\psi = \psi'$ which in practice can be approximated but not assured. The achievement of this condition is one of the goals of the mask optimization procedure described in a later section. Note that even when the condition holds, ψ still appears in the equation. Now however, it appears only as a scaling factor and presents no ambiguity if sufficiently frequent calibration is performed.

2. Gas Response and Noise

The response relationships derived in the preceding paragraph are useful in predicting the form of the calibration curve and in suggesting how the instrument performance will be affected by changes of the sky spectrum. Proper design requires also some knowledge of the noise produced in the system. The optimum design is that which provides the largest signal-to-noise ratio, S/N for the required gas detection case. S/N will here be deduced for the important case of a four-mask instrument, the response of which is given by Eq. (9). In this equation we have assumed that ψ is the same for both mask pairs. Whether or not this is true it is nevertheless necessary to measure response from some baseline which will have noise associated with it. We suppose this baseline to be the response when no gas is present as P_{oj}; P_{gj} is the corresponding quantity when a gas absorption spectrum is seen. As before we can describe the response as

$$R_o = drI_1\left[\frac{P_{o4}}{P_{o3}} - \frac{P_{o2}}{P_{o1}}\right] \quad \text{for no target gas}$$

and

$$R_g = drI_1\left[\frac{P_{g4}}{P_{g3}} - \frac{P_{g2}}{P_{g1}}\right] \quad \text{for target gas}$$

Subtracting the baseline R_o from the gas response R_g yields

$$S = drI_1\left[\left(\frac{P_{g4}}{P_{g3}} - \frac{P_{g2}}{P_{g1}}\right) - \left(\frac{P_{o4}}{P_{o3}} - \frac{P_{o2}}{P_{o1}}\right)\right] \tag{10}$$

The limiting noise in the system is assumed to be photon noise described by the Schottky equation for cathode noise current

$$i_k^2 = 2eI_k \, \Delta\nu$$

where

e is electronic charge (C)

$\Delta\nu$ is bandwidth of following electronics (Hz)

The anode current noise may be written

$$i_j^2 = cGId = c\left(P_jS_k + I_{0k}\right)G^2d$$

where c is a constant including $2e\,\Delta\nu$ and a factor dependent upon the gain of each of the dynodes.

The gain with and without gas is given by

$$G_g = \frac{I_1}{P_{g1}S_k + I_{0k}} \qquad G_o = \frac{I_1}{P_{o1}S_k + I_{0k}}$$

for pulses corresponding to masks 1 and 2 and

$$G_g = \frac{I_1}{P_{g3}S_k + I_{0k}} \qquad G_o = \frac{I_1}{P_{o3}S_k + I_{0k}}$$

for pulses 3 and 4.

The square of the noise of the final signal is the sum of the squares of the noise of each pulse with and without gas.

$$i^2 = cd\left(\frac{(P_{g1}S_k + I_{0k})I_1^2 d}{(P_{g1}S_k + I_{0k})^2} + \frac{(P_{g2}S_k - I_{0k})I_1^2 d}{(P_{g1}S_k + I_{0k})^2} + \frac{(P_{g3}S_k + I_{0k})I_1^2 d}{(P_{g3}S_k + I_{0k})^2}\right.$$
$$+ \frac{(P_{g4}S_k + I_{0k})I_1^2 d}{(P_{g3}S_k + I_{0k})^2} + \frac{(P_{o1}S_k + I_{0k})I_1^2 d}{(P_{o1}S_k + I_{0k})^2} + \frac{(P_{o2}S_k + I_{0k})I_1^2 d}{(P_{o1}S_k + I_{0k})^2}$$
$$\left.\frac{(P_{o3}S_k + I_{0k})I_1^2 d}{(P_{o3}S_k + I_{0k})^2} + \frac{(P_{o4}S_k + I_{0k})I_1^2 d}{(P_{o3}S_k + I_{0k})^2}\right)$$

Assuming again that the dark current is small we have

$$i^2 = \frac{cI_1^2 d}{S_k}\left(\frac{P_{g1} + P_{g2}}{P_{g1}^2} + \frac{P_{g3} + P_{g4}}{P_{g3}^2} + \frac{P_{o1} + P_{o2}}{P_{o1}^2} + \frac{P_{o3} + P_{o4}}{P_{o3}^2}\right)$$

The noise in rms volts is $n = r\sqrt{i^2}$

$$n = \left(\frac{cI_1^2 d}{S_k}\right)^{1/2}\left(\frac{P_{g1} + P_{g2}}{P_{g1}^2} \quad \frac{P_{g3} + P_{g4}}{P_g3^2} \quad \frac{P_{o1} + P_{o2}}{P_{o1}^2} \quad \frac{P_{o3} + P_{o4}}{P_{o3}^2}\right)^{1/2} \tag{11}$$

Finally, the signal-to-noise ratio is from Eqs. (10) and (11):

$$S/N = \left(\frac{S_k d}{c}\right)^{1/2} \frac{\left[\left(\frac{P_{g4}}{P_{g3}} - \frac{P_{g2}}{P_{g1}}\right) - \left(\frac{P_{o4}}{P_{o3}} - \frac{P_{o2}}{P_{o1}}\right)\right]}{\left(\frac{P_{g1} + P_{g2}}{P_{g1}^2} + \frac{P_{g3} + P_{g4}}{P_{g3}^2} + \frac{P_{o1} + P_{o2}}{P_{o1}^2} + \frac{P_{o3} + P_{o4}}{P_{o3}^2}\right)^{1/2}} \tag{12}$$

The equation states that the signal-to-noise ratio is proportional to the square root of the duty cycle d and the photomultiplier sensitivity S_k. A common factor $A\Omega$ can be extracted from each of the P_j which would show the variation of S/N with the square root of the throughput as well.

Had Eq. (12) been derived without the assumption of negligible dark current an inverse dependence on dark current would have resulted. This indicates that the AGC controlling current must be large enough to guarantee the condition which in turn requires that the masks be designed to keep the pulses as large as possible.

3. Alternate Signal Processing

The analysis presented applies to all correlation spectrometers so far produced by BRL. Recently an alternate method of signal processing has been derived that shows promise of more flexibility in preventing ambiguity caused by sky changes. The second AGC is eliminated and each of the four pulses corresponding to the four mask positions is provided with its own selected gain. The response, following the same analysis as in Sec. II.B.1 and describing the individual gains by b_i, is

$$V = rdI_1 \frac{[b_1P_1 - b_2P_2 - b_3P_3 + b_4P_4]}{P_1} \tag{13}$$

Again introducing average absorption coefficients and extending the ψ to mean the same ratio as before but always relative to mask 1 leads to the final response equation

$$V = rI_1d\{b_1 - b_2\psi \exp[(a_1 - a_2)cL] - b_3\psi' \exp[(a_1 - a_3)cL] + b_4\psi'' \exp[(a_1 - a_4)cL]\} \tag{14}$$

Without evaluating the b_i or the ψ^j we see from this expression that the form of response curve is again the algebraic sum of exponentials. The important difference between Eq. (13) and Eq. (9) is that we now have several extra variables that we can use to cancel out certain changes in ψ^j. If no gas is present, Eq. (13) shows that changes of any of the ψ^j can produce a voltage response. Differentiating with respect to time and setting $dV/dt = 0$ gives

$$- b_2\dot{\psi} - b_3\dot{\psi}' + b_4\dot{\psi}'' = 0$$

If now the b_j are chosen according to

$$b_j \propto \frac{1}{\dot{\psi}^j}$$

then the sky changes described by $\dot{\psi}^j$ are ineffective in changing the response.

In order to find appropriate values of b, it is useful to consider the mathematical form of any expected spectral changes. The most obvious change is one of light level without a change of spectral shape, i.e., multiplication of the spectrum $N(\lambda)$ by a constant m. This type of change is already compensated by the AGC in this signal processing method. A second possibility is the superposition of a flat continuum from an independent source or the addition of a constant C to every point of the spectrum. Such a change could be compensated by the signal processing, which takes differences between parts of the spectrum, but the action of the AGC prevents this. By choosing the b_j carefully, however, both additive and multiplicative changes of $N(\lambda)$ can be excluded from the response. To find these particular b_j we write the bracketed factor in Eq. (13) for cL = 0.

$$b_1P_1 - b_2P_2 - b_3P_3 + b_4P_4 = S$$

The spectral change now provides new values P'_j but leaves S unchanged

$$b_1P'_1 - b_2P'_2 - b_3P'_3 + b_4P'_4 = S \tag{15}$$

where

$$P'_j = A\Omega \iint_{-\infty}^{\infty} [mN(\lambda) + C]\beta(\lambda)F_j(\lambda - \xi)\ d\lambda\ d\xi$$

according to Eq. (4) and the supposed conditions. Further

$$P'_j = mP_j + C \iint_{-\infty}^{\infty} A\Omega\beta(\lambda)F_j(\lambda - \xi)\ d\lambda\ d\xi = mP_j + Cwj$$

where w_j is defined as the double integral. Equation (15) is thus

$$(b_1P_1 + b_2P_2 - b_3P_3 + b_4P_4)m + (b_1w_1 - b_2w_2 - b_3w_3 + b_4w_4)C = S$$

This expression is insensitive to changes in both m and C if

$$b_1P_1 - b_2P_2 - b_3P_3 + b_4P_4 = 0$$

and

$$b_1w_1 - b_2w_2 - b_3w_3 + b_4w_4 = 0$$

The first of this pair states that S = 0 which means the response for zero gas must be set to zero volts. In practice b_1 is a redundant parameter since it merely changes the AGC gain factor. We therefore let $b_1 = 1$. In the four-mask instruments presently in use the second and third masks are identical, so that forcing $b_2 = b_3$ causes no loss of generality. Thus we have a pair of simultaneous equations from which $b_2 = b_3$ and b_4 can be calculated.

Another possible type of sky change is a change of the spectral slope

$$N'(\lambda) = N(\lambda)(f\lambda + h)$$

The bracketed factor of Eq. (13) then leads to

$$(b_1Q_1 - b_2Q_2 - b_3Q_3 + b_4Q_4)f + (b_1P_1 - b_2P_2 - b_3P_3 + b_4P_4)h = S$$

where

$$Q_j = \iint A\Omega\lambda N(\lambda)\beta(\lambda)F_j(\lambda - \xi)\ d\lambda\ d\xi$$

Again this expression is insensitive to any values of f and h if

$$b_1Q_1 - b_2Q_2 - b_3Q_3 + b_4Q_4 = 0$$

and

$$b_1P_1 - b_2P_2 - b_3P_3 + b_4P_4 = 0$$

With the real conditions that $b_1 = 1$ and $b_2 = b_3$, the pair of equations can be solved simultaneously to give b_2 and b_4 that make the response insensitive to slope changes of the form assumed above.

Making the bracketed factor in Eq. (13) insensitive to particular types of sky changes is best accomplished by letting it be zero for zero gas. A change of $N(\lambda)$ when gas is present will also affect the denominator of Eq. (13), but it does so as a scaling factor. As in the previous signal processing method, this presents no difficulty because frequent calibration provides suitable correction.

The two examples considered here involve only linear changes of the sky spectrum, i.e., those which require only two independent parameters for complete specification. Changes of spectral curvature would require at least three such parameters and four masks, no two of which are identical.

A quantitative empirical determination of b_j can be performed by considering Eq. (14). It is just the algebraic sum of four voltages which can be adjusted at will by the four adjustable gains b_j. If we view the sky at a time when there is no target gas present we have

$$V = b_1V_1 - b_2V_2 - b_3V_3 + b_4V_4 = 0$$

At some later time we suppose a sky change has occurred but still no gas is present, and we have

$$V' = b_1V_1' - b_2V_2' - b_3V_3' + b_4V_4' = 0$$

Measuring V_j and V_j' and using $b_1 = 1$ and $b_2 = b_3$ permits calculation of b_2 and b_4 which make the instrument insensitive to the kind of change which has occurred. Only if the change affected the spectrum in a linear fashion will these b_j values apply to any other observations, however.

The signal-to-noise ratio for these signal processing methods is found by the same procedure as shown in Sec. II.B.2. The signal is again the difference between the gas signal and the baseline, even though the baseline is at zero volts.

$$S = I_1 rd\left(\frac{b_1P_{g1} - b_2P_{g2} - b_3P_{g3} + b_4P_{g4}}{P_{g1}} - \frac{b_1P_{o1} - b_2P_{o2} - b_3P_{o3} + b_4P_{o4}}{P_{o1}}\right)$$

The noise from the cathode is now multiplied by both the photomultiplier gain and the b_j

$$i_j = Gb_j(2e\ \Delta\nu I_k)^{1/2}$$

or

$$i^2 = \left(\frac{cdI_1^2}{S_k}\frac{b_1^2P_{g1} + b_2^2P_{g2} + b_3^2P_{g3} + b_4^2P_{g4}}{P_{g1}^2} + \frac{b_1^2P_{o1} + b_2^2P_{o2} + b_3^2P_{o3} + b_4^2P_{o4}}{P_{o1}^2}\right)$$

The noise present in the response is again $n = ri$ and

$$S/N = \left(\frac{S_k d}{c}\right)^{1/2}\left(\frac{b_1P_{g1} - b_2P_{g2} - b_3P_{g3} + b_4P_{g4}}{P_{g1}} - \frac{b_1P_{o1} - b_2P_{o2} - b_3P_{o3} + b_4P_{o4}}{P_{o1}}\right) \Bigg/ \left(\frac{b_1^2P_g + b_2^2P_{g2} + b_3^2P_{g3} + b_4^2P_{g4}}{P_{g1}^2} + \frac{b_1^2P_{o1} + b_2^2P_{o2} + b_3^2P_{o3} + b_4^2P_{o4}}{P_{o1}^2}\right)^{1/2} \qquad (16)$$

Again the S/N depends upon the square root of the duty cycle, cathode sensitivity, and throughput (via P_j).

C. Mask Optimization

The analysis of Sec. II.B assumed that some sort of masks were available, but only a brief mention was made of the proper design of the mask. Early masks were constructed by photographing the spectrum of the target gas in the exit plane of the spectrometer. The photographic negative was used directly, transmitting in the absorption bands and blocking the light elsewhere. However, such a mask is only one of many possible designs, and it is important to select a design which will meet specific needs. For most purposes, particularly in the ultraviolet using the sky as a spectral source, the main consideration is a large S/N. Another consideration may be a desired insensitivity to interference. The masks in use at BRL were designed by Millan (1972) for maximum signal-to-noise ratio and it is his procedure which is described in what follows.

The optimization of a mask set for a four-mask correlation spectrometer ideally consists of differentiating Eq. (12) with respect to every possible parameter and setting each partial derivative equal to zero. The resulting set of simultaneous equations may then be solved for each of these parameters separately. Of course, such a procedure is cumbersome and some simplifications are in order. The first is that only one mask pair is optimized at a time.

The S/N for a two-mask instrument is

$$S/N = \left(\frac{dS_k}{c}\right)^{1/2} \frac{\left[\dfrac{P_{o2}}{P_{o1}} - \dfrac{P_{g2}}{P_{g1}}\right]}{\left(\dfrac{P_{o1} + P_{o2}}{P_{o1}^2} + \dfrac{P_{g1} + P_{g2}}{P_{g1}^2}\right)^{1/2}} \qquad (17)$$

This expression is maximized for the three slit parameters ξ (slit position), Δ (slit width), and δ (mask separation) for the simple case of two single-slit masks looking at a sinusoidal absorption spectrum imposed on a spectrally flat background as shown in Fig. 6.

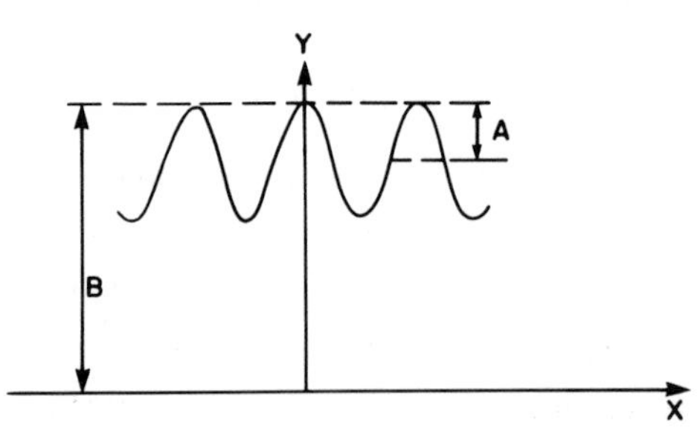

FIG. 6. Two single-slit masks looking at sinusoidal absorption spectrum imposed on spectrally flat background.

The results of maximization are that the absorption at either edge of a slit should be the same, that the slit should be of width 2.33 radians, and that the entrance and exit slits should be of equal width. These conditions are taken as starting parameters for the real target gas spectrum as measured in the laboratory with a high resolution scanning spectrometer. This spectrum is multiplied by an assumed sky spectrum $N(\lambda)$ and the sensitivity function of the detector. The starting parameters are used to arrive at a value of S/N by Eq. (17) and then changed one angstrom at a time to see whether a higher S/N can be found. In this way the combination of slit parameters giving maximum signal-to-noise ratio for each of the bands for the case of a single-slit mask pair is ascertained.

The next stage of optimization is to determine the number of exit slits, n, that maximize S/N. Millan (1972) showed by a rather long mathematical argument involving some fairly restrictive approximations that the maximization conditions

$$\frac{\partial(S/N)}{\partial x_i} = 0$$

lead to 3n simultaneous equations that are uncoupled in the sense that each slit acts independently of all the others. Therefore the S/N for each absorption band can be used in the following manner. The slit pair giving the maximum S/N and the slit pair giving the second largest are combined into a double slit mask pair, the S/N of which is evaluated by Eq. (12). If the combined S/N is greater than that of either one-slit pair it is known that at least two slits are required for each mask. The slit pair with the third largest S/N is added to the masks and the S/N by Eq. (12) is compared with the double slit value. This procedure continues until

$$(S/N)_i \leq (S/N)_{i-1}$$

When the S/N decreases with the addition of another slit pair the optimum number of slits has just been exceeded. An approximately optimized mask pair is thus designed.

It is still necessary to design the second mask pair if a four-mask instrument is desired. The most important consideration for the second mask according to Sec. II.B.1 is that $\psi_1 = \psi_2$. This condition can be satisfied by careful regard to the background spectrum and its anticipated variations. For a gas with a periodic band structure in a region of uncomplicated sky spectrum, the simplest procedure is to make the second mask pair identical to the first (optimum) pair and displace it by just one band. Some trial and error calculations of Eq. (12) on the second mask pair could be helpful.

It must be emphasized that the mask design at this stage is only approximate. The full S/N Eq. (12) does not lead to uncoupled solutions upon maximization. With the approximate parameters taken as starting values Eq. (12) may be optimized by a trial and error procedure. This final task must include the effect of a filter to exclude scattered light from within the spectrometer, i.e., the complete $\beta(\lambda)$. It should begin with a single entrance slit and include a procedure for deciding whether one, three, or five entrance slits are optimum, again by trial and error. The positions of the side entrance slits relative to the central slit are such that the spectral pattern produced by each side slit coincides with that of the central slit displaced by one band.

D. Instrumentation

1. Gas-Sampling Spectrometer with Active Light Sources

It has already been stated in Sec. I that the response of a correlation spectrometer, as in any absorption spectrometer, depends upon the optical thickness of the target gas. This was written in Eqs. (8) and (9) as cL with the understanding that the measurable quantity is $\int c\,dL$. Only if the variation of c with L is known can the concentration c be determined directly. An instrument operating in the fixed path or active mode satisfies this condition by supplying a known L and measuring a value c averaged over this distance. A light source of constant spectral composition is an essential feature of such a device, and this feature eliminates the need for more than one mask pair.

The simplest application of dispersive correlation spectroscopy is the ambient monitor. In this model the first mask is the entrance slit array, and the second, identical to the first, is the exit plane. Both are stationary. A torque motor imparts a rotational oscillation to the grating to produce a modulated signal at the photomultiplier. The response depends upon the concentration of gas inside the absorption cell according to the two-mask response Eq. (8). Using a path length of 2.5 m and signal integration time of 100 sec the noise level in SO_2 and NO_2 measurements is around 15 ppb.

A second fixed-path correlation spectrometer is the in-stack monitor produced by Combustion Engineering Associates, Inc. and shown schematically in Fig. 7. Light from a quartz-iodine lamp is projected across part of the smokestack to a mirror which reflects it back into the correlation spectrometer. In this instrument also the two stationary masks are at the entrance and exit planes of the spectrometer. The spectrum oscillates across the exit mask because of reflection from a rotating slanted mirror. The masks in this instrument are photographic reproductions of the SO_2 spectrogram. A correlation type in-stack monitor is also produced by Environmental Data Corporation.

2. Remote Sensing Spectrometers

The most successful application of dispersive correlation spectroscopy so far is the monitoring of air pollution using the sky as the light source. In this application the variation of the target gas concentration along the absorption path is not generally known, but the total vertical burden, $\int c dl$ (in units of ppm-m or atm-μm) is measured by pointing the instrument vertically. This quantity is particularly useful in determining the mass flow of pollutant across a traverse line along which the spectrometer is transported if the wind direction and speed are known.

The first model of the remote sensing correlation spectrometer built at Barringer Research is shown in Fig. 8. It consisted of an Ebert spectrometer with a single entrance slit and a single stationary mask of multiple exit slits. A pair of refractor plates joined

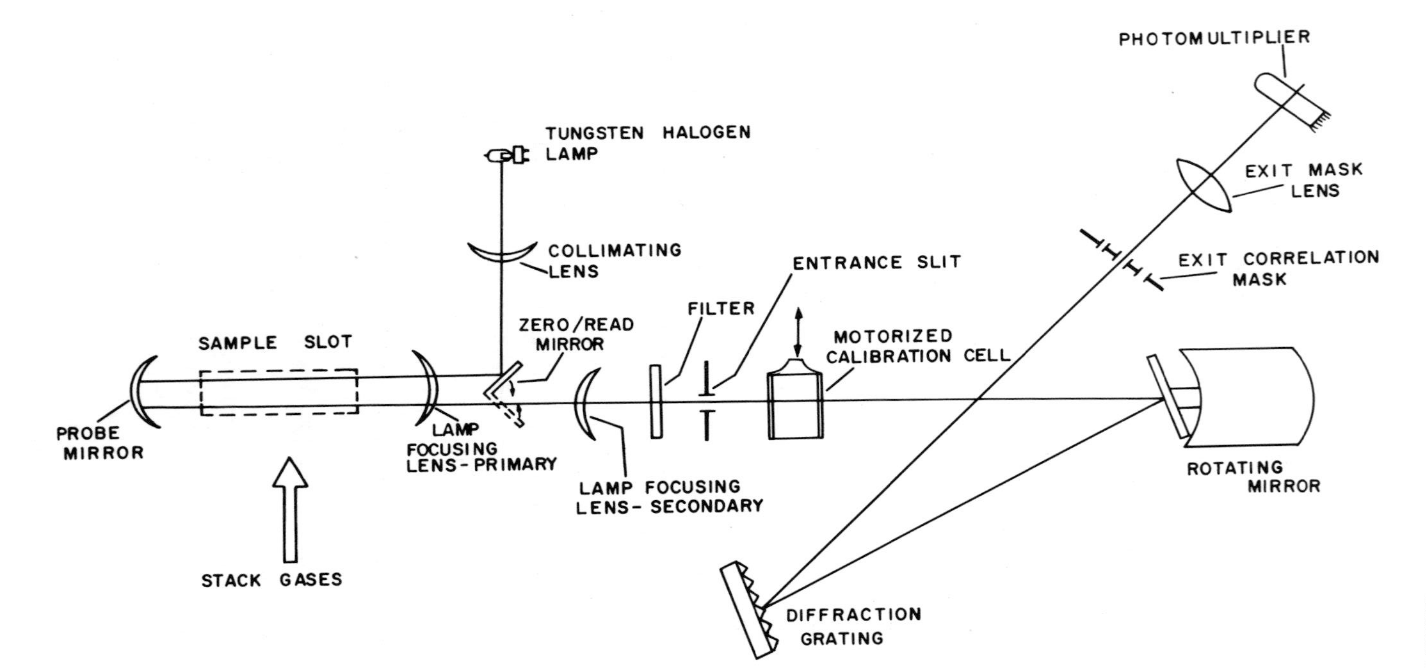

FIG. 7. Scheme of in-stack monitor (fixed-path correlation spectrometer).

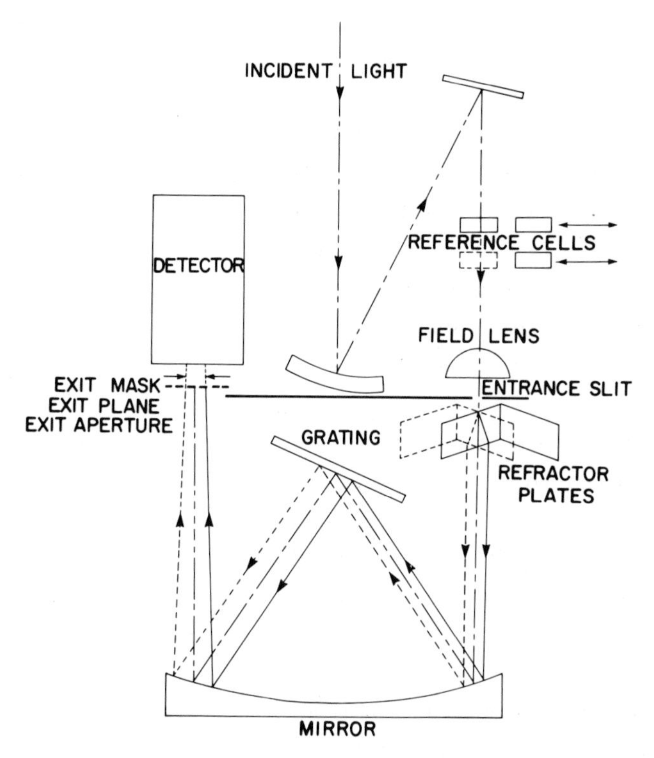

FIG. 8. Model of remote sensing correlation spectrometer. (Millan and Newcomb, 1970)

at a corner vibrates behind the entrance slit in such a way that the image of the entrance slit, and hence the spectrum, shifts by a fixed amount each time the corner of the refractor plate assembly crosses the light path. The spectrum alternates between two positions along the exit mask. The first position matches the transmission maxima with the slits; the second position matches the minima with the slits. The operation is described by all the analysis of Sec. II.B that pertains to a two mask instrument. In this case the two masks are of course identical and the distance between the masks, δ, is actually the distance the spectrum jumps when the refractor plate junction crosses the beam. Because of the two positions of the spectrum the instrument is known as a bistable correlation spectrometer. Its disadvantage is that its response is described by Eq. (8) with its consequent ambiguity of baseline changes. That is, changes of background sky spectrum can cause changes in ψ which in turn add an offset to the response.

An attempt to overcome the offset problem inherent in the bistable spectrometer was the bistable scanning spectrometer described by Millan and Newcomb (1970). The mechanical operation is the same as that of the bistable instrument shown in Fig. 8 except that a slow grating rotation is added to the faster slit oscillation motion. The spectrum moves from a position where the jump is from maxima to minima of correlation and back, through a position where the transmission is the same at both ends of the jump, to a position where the jump is from minima to maxima of correlation and back. The gas response term of Eq. (8) thus goes from positive values ($a_1 > a_2$) through zero ($a_2 = a_1$) to negative values ($a_2 > a_1$). If ψ remains constant over the spectral region of the scan, the response is an oscillation of the gas term about the zero gas term. An example of a record from such an instrument is reproduced in Fig. 9. Although a moving baseline is still possible, only the amplitude of the oscillating response is needed to determine cL, since the scanning time is generally short compared to anticipated ψ changes. Unfortunately this method does not completely eliminate the dependence on ψ in the zero gas term if the background spectrum is not smooth over the region of spectral scanning. If a spectral feature is present in the background, ψ also oscillates with the same frequency as the gas

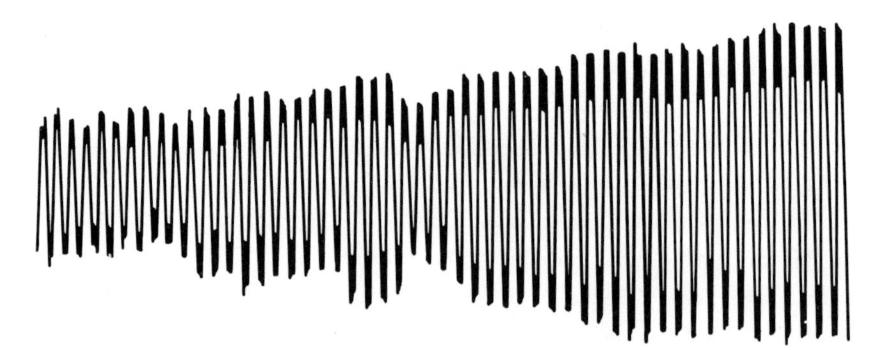

FIG. 9. Recording from remote sensing correlation spectrometer. (Millan and Newcomb, 1970)

term and hence the zero gas term adds algebraically to the amplitude. This problem notwithstanding, the bistable scanning technique represents a significant improvement over the bistable. Neither technique is insensitive to slope or curvature changes of the sky spectrum.

The present series of correlation spectrometer produced at Barringer Research makes use of the four-mask advantage discussed in conjunction with Eq. (9). The four masks are etched on a rotating disk so that they pass through the light beam in the exit focal plane sequentially. This instrument is described in detail in the following section.

3. The Present Barringer Correlation Spectrometer or COSPEC

a. Optical. The correlation spectrometer presently built and marketed by Barringer Research under the trade name COSPEC is shown in the diagram of Fig. 10. The spectrometer itself is a quarter-meter f/4 Ebert Fastie system, optimized for minimum aberration and maximum resolution. The blazed reflection grating is produced by Jarrel-Ash Corp. with 1180 lines/mm. Light is collected by a Cassegrain telescope with a field of view 10 mrad and focused on three curved entrance slits etched photographically on an aluminum plated quartz slab. Forty-five degree plane mirrors placed next to the entrance and exit slits allow an orientation of the spectrometer perpendicular to the viewing direction. The dispersed light leaves the spectrometer housing through a baffling port, passes through the exit masks and a broad filter, and strikes the photomultiplier tube.

The correlation masks are photoetched in aluminum on a quartz disc according to the design procedure of Sec. II.C. A mask for SO_2 detection is shown in Fig. 11. The axis of the rotating disc is optically at the center of the Ebert mirror so that the slit curvature is that required by the spectrometer design. The slit function $F(\lambda)$ for SO_2 is shown with the SO_2 absorption spectrum in Fig. 12. A Corning 9-54 filter absorbs visible light that passes through the exit plane.

The correlation disk rotates at slightly less than 60 Hz. As it turns, light from an array of photoemitting diodes passes through

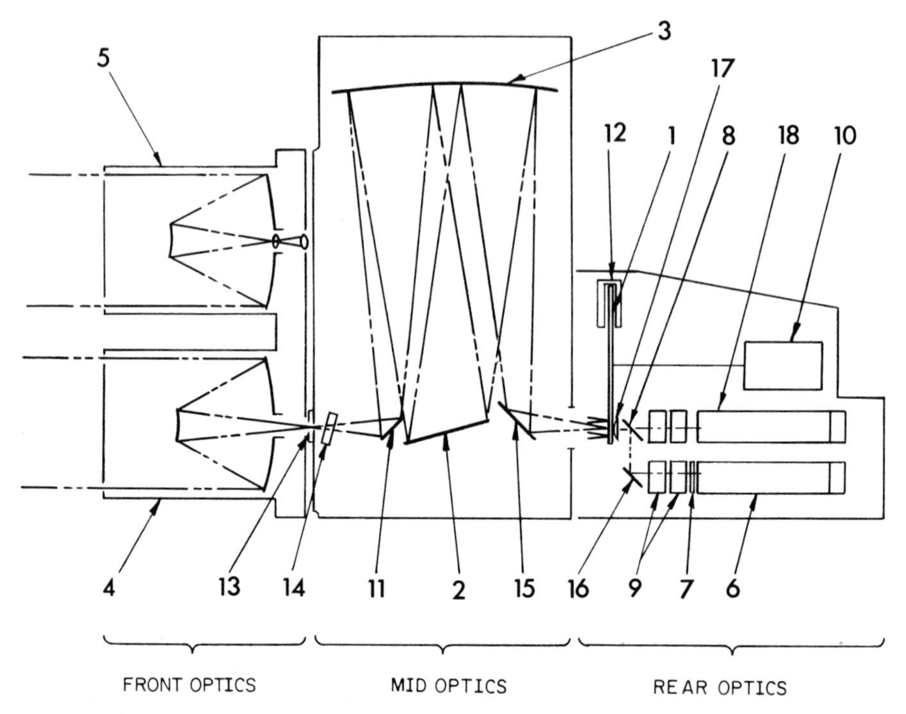

FIG. 10. COSPEC II ray diagram. 1, Chopper disc; 2, gratings; 3, spherical mirror; 4, direct record Cassegrain; 5, Maxwellian view Cassegrain; 6, SO_2 photomultiplier tube; 7, filter; 8, dichroic mirror; 9, calibration cells; 10, motor; 11, mirror; 12, logic diodes; 13, entrance slit; 14, refractor plate; 15, mirror; 16, mirror; 17, prism; 18, NO_2 photomultiplier tube.

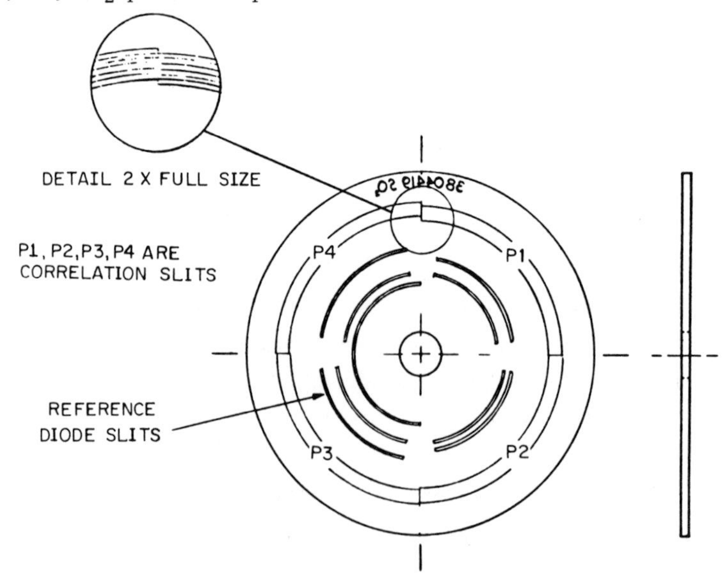

FIG. 11. COSPEC IV mask for SO_2 detection.

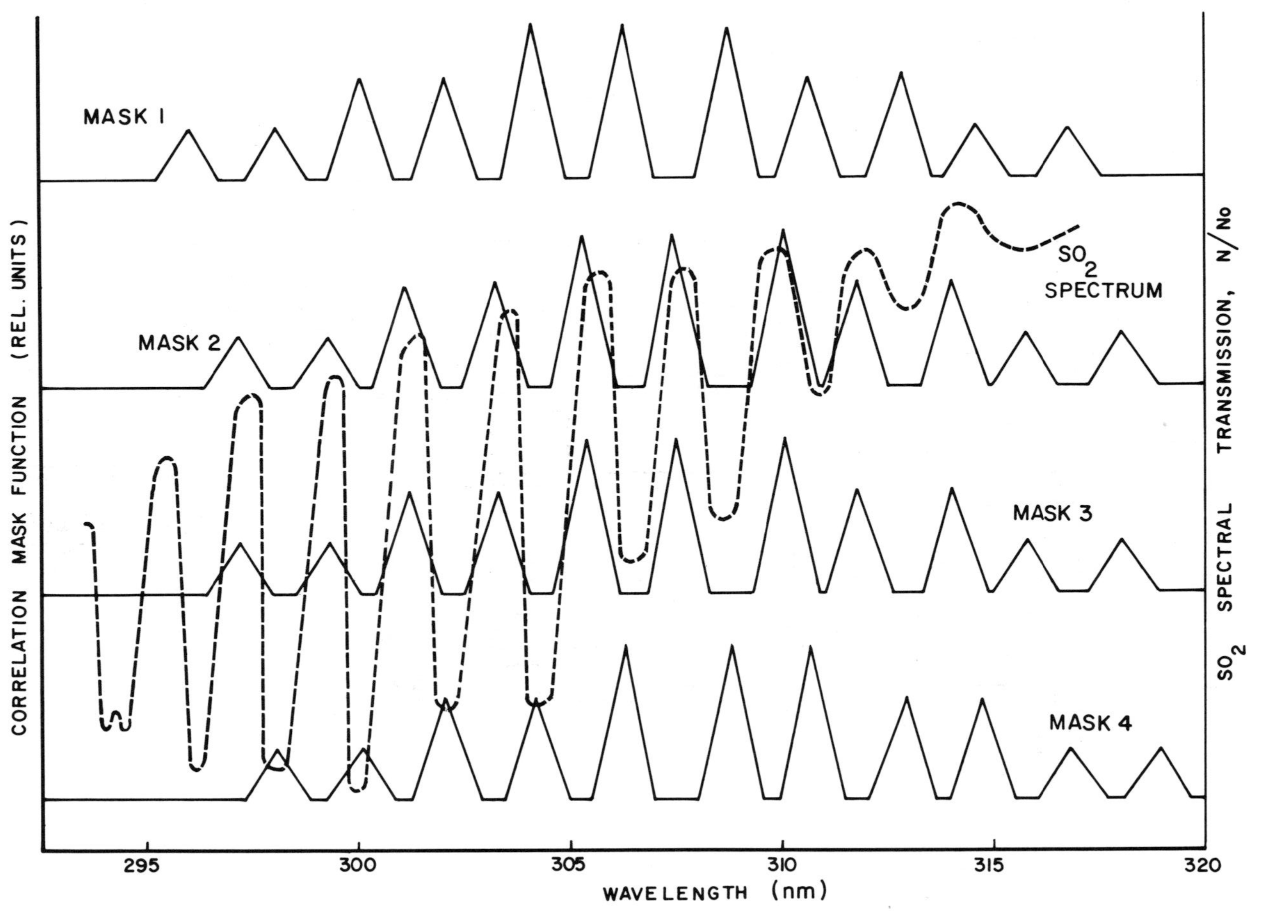

FIG. 12. Correlation mask for SO_2 detection.

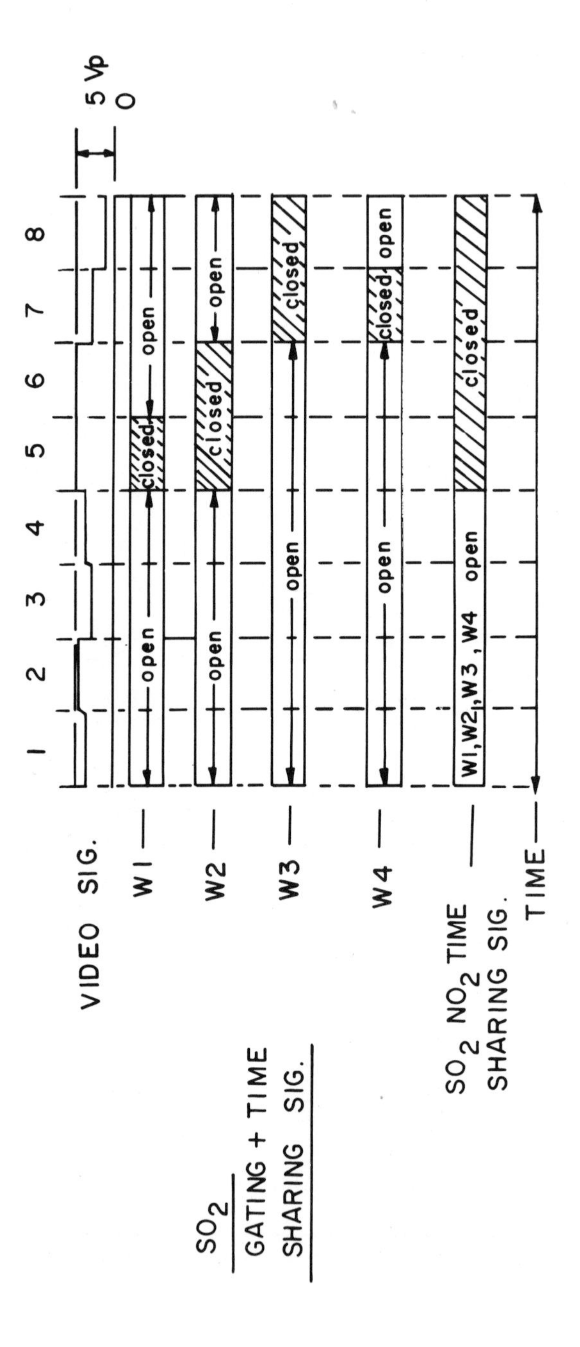
SO2
GATING + TIME
SHARING SIG.
VIDEO SIG.
W1
W2
W3
W4
SO2 NO2 TIME
SHARING SIG.
TIME
1
2
3
4
5
6
7
8
5 Vp
0
open
closed
W1, W2, W3, W4 open

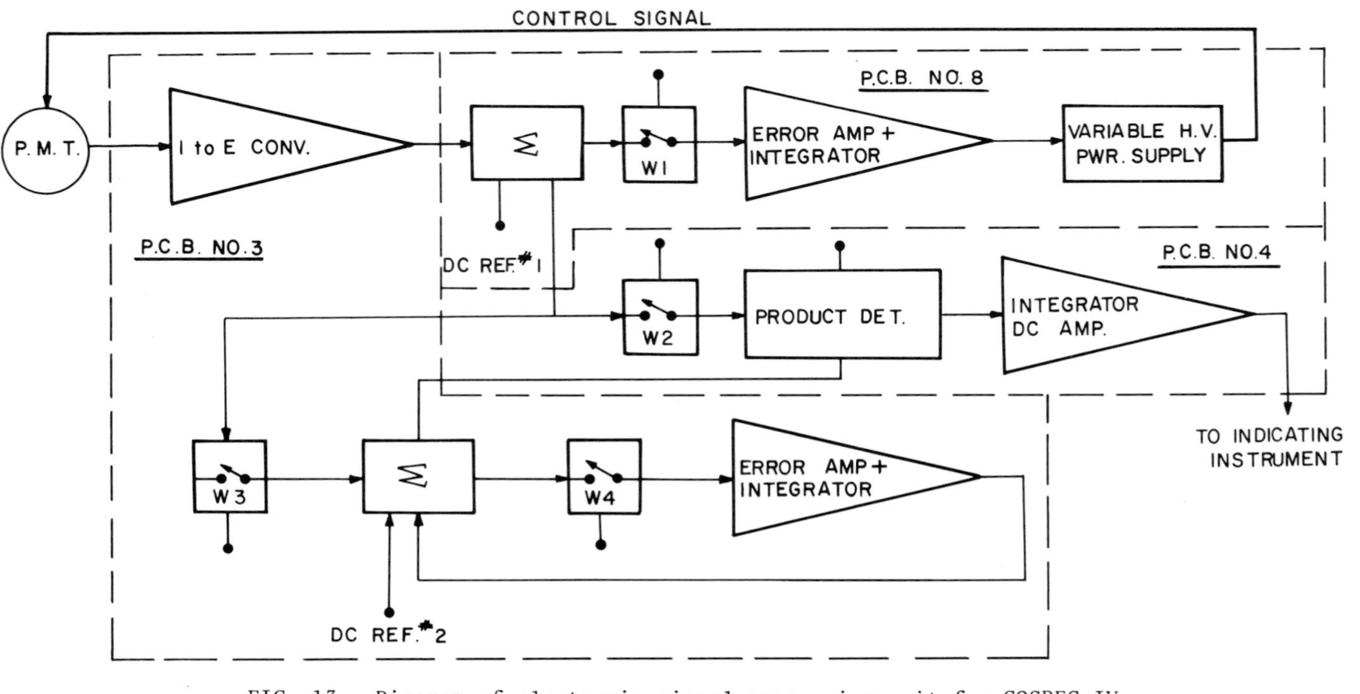

FIG. 13. Diagram of electronic signal processing unit for COSPEC IV.

an inner bank of wide slits for detection by an array of corresponding phototransistors which generate timing signals for the electronic processing unit.

b. Electronic. A block diagram of the electronic signal processing unit is given in Fig. 13. The current at the anode of the Phillips XP1118 photomultiplier consists of four pulses, one for each mask position of the rotating disc. The pulse train is amplified and converted to a series of voltage pulses, the first of which is compared with a preset dc reference at the summing junction Σ. Any voltage difference between the reference and the first pulse is passed through switch W1 to the error amplifier which adjusts the high voltage supply to bring the error signal back to zero. Switch W1 is controlled by the timing pulses from the inner band of the correlation disc to be closed only when the first pulse is being generated so that the PMT gain remains the same when all but the first pulse enter the system. In this way pulse one is always maintained at the preset level with all other pulses remaining in the same ratio to pulse one as their total radiant powers. Both pulses 1 and 2 pass through switch W2 to enter the product detector. Switch W2 opens and switch W3 closes for pulses 3 and 4. When pulse 3 arrives at the summing junction it is with another preset dc reference and the difference is passed through W4 for amplification. The output of the error amplifier adjusts the gain at the second summing junction to give zero error and W4 opens, leaving the gain as adjusted for the passage of pulse 4. The pulse time is reconstructed at the product detector and passed on to the integrator. Before integration the second and third pulses are inverted, i.e., multiplied by -1. The integrator then sums all four pulses which now means

$$V = V_1 - V_2 - V_3 + V_4$$

The integrated voltage is amplified and registered by a chart recorder.

c. Variations. The instrument just described, called the COSPEC IV, is designed for the measurement of a single target gas when the available light level is low. An example is SO_2 measurements at 3000 Å in winter at high latitudes. However, the correlator disc makes a dual gas instrument possible when light requirements can be relaxed. The information on the disc of the COSPEC IV is compressed to cover only half a disc and a second set of four masks appropriate to the second gas is etched on the other half of the disc. The timing bands are repeated and a new band is added to provide a switch to separate the signals from the two gases. The second gas also requires a different grating position in general. For this reason two half gratings are used, one above the other. The light beam is thereby split into two parts at the grating and the parts are brought back together at the exit plane. A dichroic mirror is used to separate the two spectra after passage through the correlator disc and to pass them to separate photomultipliers. Two identical electronic processing units operate in parallel, one for either gas on alternate half-cycles. The second gas is usually NO_2 in the wavelength range of 4300 Å. The instrument is called the COSPEC II. Since either gas is now viewed by only half the grating area of the COSPEC IV, the throughput $A\Omega$ is reduced and consequently the signal-to-noise ratio as well. Moreover, the duty cycle in this time-sharing configuration is only half that of the COSPEC IV and again the signal-to-noise ratio is reduced.

A COSPEC III has been produced which is identical to the COSPEC II when used as a passive remote sensor. It offers the additional possibility of observations on a modulated xenon arc lamp, that is, active mode operation. A collimated beam from the xenon lamp is transmitted up to 1000 m to the COSPEC. Modulation of the light source at 2.5 kHz and phase sensitive detection of the modulated signal in the COSPEC assures that only light from the lamp is used in the measurement without interference by sunlight.

III. GAS FILTER CORRELATION SPECTROSCOPY

A. General

The technique of gas filter correlation spectroscopy utilizes the difference in transmission between a cell containing a sample of the target gas and an identical cell containing no target gas. A spectrometer based on this technique is a correlation device in the sense that the concentration of the target gas in the air to be analyzed is measured by the degree of correlation between its spectrum and that contained within the sample cell of the instrument.

The absorption spectra of the sample and the reference cells for a spectrally flat, continuous source are shown in the top of Fig. 14 for a hypothetical target gas. When a small amount of target gas is introduced between the light source and the sensor the absorption spectra seen behind the cells are those in the center of Fig. 14. The change of transmission through the sample cell, comparing the top and center portions of Fig. 14, is small, but that through the reference cell is substantial. In other words, the sample cell is opaque to the absorption spectrum of the target gas and hence cannot transmit changes in the depth of this spectrum. If a suitable detector alternately views the light source through the sample cell and the reference cell a modulated signal is produced, the modulation being deeper the more target gas there is in front of the sensor. Normally a gain adjustment is applied to the sample cell signal to set the modulation to zero for zero target gas. The modulation then increases with increasing target gas concentration.

A low-resolution technique is often used for the same purpose and in the same way. An interference filter is substituted for the sample cell, selectively absorbing the whole spectral region shown in Fig. 14. Using a proper source and a spectrally pure target gas the technique produces the same type of modulation as the gas filter method. However, if a gas with an interfering spectrum is introduced into the target gas as in the bottom of Fig. 14, the interference filter will also be opaque to the interferent and produce a gas type modulation. The gas filter correlation spectrometer, on the

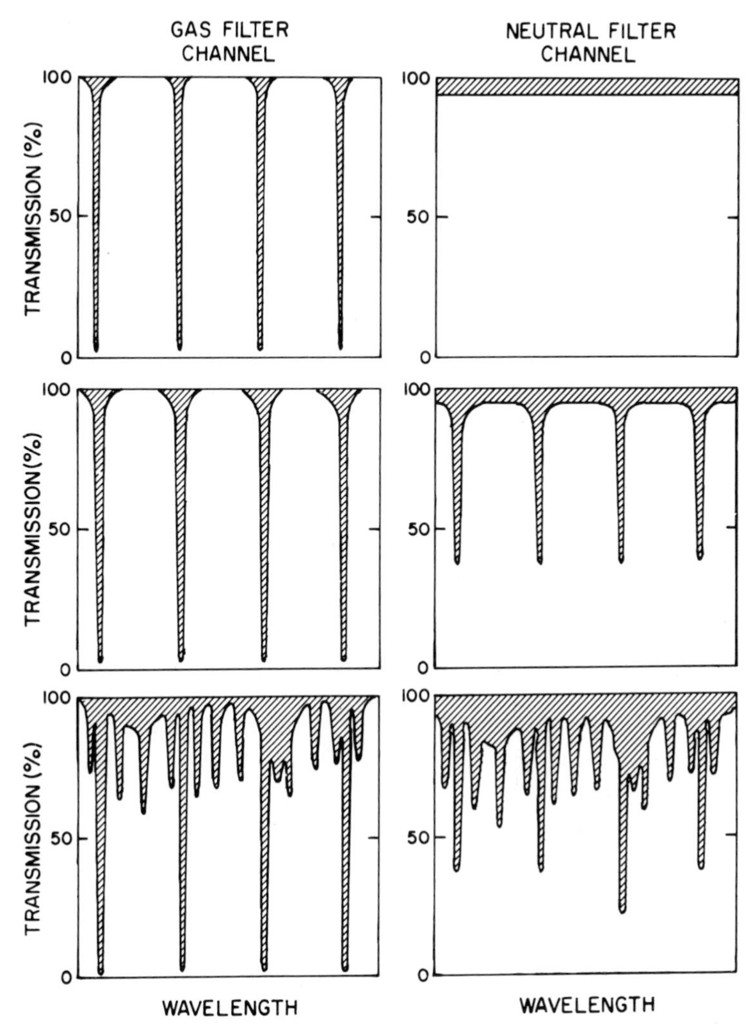

FIG. 14. Principles of gas filter correlation spectrometers. (Hodgeson et al., 1973)

other hand, sees the interference in both channels so that no change in modulation is produced. The gas filter correlation technique is thus more specific than the filter technique due to its higher resolution correlation. Gas filter correlation spectrometers have been constructed for both active and passive mode analysis, usually for measurement in the infrared region of the spectrum. As indicated in

Sec. I.B passive mode observations in the infrared often use the emission spectrum of the target gas. If the target gas itself constitutes the source, the sample cell channel is again insensitive to changes of the spectrum, but the reference channel shows an increasing signal for increasing target gas concentration. Modulation is now in the opposite sense to the absorption case.

1. Active Mode Sensing

The spectral radiant power available to the detector after passing through the sample cell is

$$p_1(\lambda) = A\Omega \int_0^\infty \{N_1(\lambda, T_s) \exp[-k(\lambda)cL] - N_2(\lambda, T_c)\} \beta(\lambda)\tau_1(\lambda)\, d\lambda \tag{18}$$

where

$A\Omega$ is instrumental throughout ($cm^2 \cdot sr$)
$N_1(\lambda, T_s)$ is source radiance ($W \cdot cm^{-2} \cdot sr^{-1} \cdot \mu m^{-1}$)
$N_2(\lambda, T_c)$ is radiance from dark side of chopper ($W \cdot cm^{-2} \cdot sr^{-1} \cdot \mu m^{-1}$)
$\beta(\lambda)$ is optical transfer function of instrument excluding sample cell
$\tau_1(\lambda)$ is transmittance of sample cell
$k(\lambda)$ is absorption coefficient of the target gas [$(atm\text{-}cm)^{-1}$ or $(ppm\text{-}m)^{-1}$]

The spectral radiant power available to the detector after passing through the reference cell is

$$p_2(\lambda) = A\Omega \int_0^\infty \{N_1(\lambda, T_s) \exp[-k(\lambda)cL] - N_2(\lambda, T_c)\} \beta(\lambda)\tau_2(\lambda)\, d\lambda \tag{19}$$

These optical signals are converted to voltages by the detector responsivity and gain factors R_1 and R_2 and the difference is taken.

$$\Delta V = R_1 p_1(\lambda) - R_2 p_2(\lambda) = S$$

$$S = A\Omega \{\int_0^\infty N_1(\lambda) \exp[-k(\lambda)cL] - N_2(\lambda, T_c)\} \beta(\lambda)(R_1\tau_1 - R_2\tau_2)\, d\lambda \tag{20}$$

To make $\Delta V = 0$ for $cL = 0$, R_2 is adjusted via its preamplifier according to

$$R_2 = \int R_1 \tau_1(\lambda)\, d\lambda / \int \tau_2(\lambda)\, d\lambda = R_1 \bar{\tau} \tag{21}$$

For active mode measurements the only unknown in Eq. (20) is cL.

For a constant source and stable instrument the integral of Eq. (20) is a negative exponential function of cL. If changes of source intensity $N_1(\lambda)$ occur they change S by a scale factor which can be corrected by frequent calibration. Changes of $N_1(\lambda)$ can also be measured directly by monitoring V_1 ($V_1 = R_1 p_1(\lambda)$) or eliminated by forming the ratio S/V_1.

The noise superposed on either value of V is assumed to be detector noise only, for which the equivalent radiant power is

$$\text{NEP} = \frac{\sqrt{A_D\ \Delta\nu}}{D^*_\lambda}$$

where

A_D is detector area (cm^2)

$\Delta\nu$ is bandwidth of processing electronics (Hz)

D^*_λ is figure of merit of the detector $W^{-1} \cdot cm^{1/2} \cdot Hz^{1/2}$)

The actual signal includes the noise from either channel for both gas and zero-gas conditions so that the signal-to-noise ratio is

$$S/N = \frac{A\Omega D^*_\lambda \int_{-\infty}^{\infty} \{N_1(\lambda, T_s)\ \exp[-k(\lambda)cL] - N_2(\lambda, T_c)\} \beta(\lambda)(\tau_1 - \bar{\tau}\tau_2) R_1\ d\lambda}{\sqrt{2}\ \sqrt{A_D\ \Delta\nu}} \tag{22}$$

2. Passive Mode Sensing

In Sec. I.C.2 it was shown that the spectral radiance $N(\lambda)$ from a source of radiance $N_o(\lambda)$ after passing through an isothermal, non-scattering atmosphere of transmittance t is

$$N_1(\lambda) = N_o(\lambda)t(\lambda) + [1 - t(\lambda)]B(\lambda, T)$$

or

$$N_1(\lambda) = t_g(\lambda)\varepsilon_b(\lambda)B(\lambda, T_b) + \varepsilon_g(\lambda)B(\lambda, T_g) \tag{23}$$

where $B(\lambda, T)$ is the Planck radiance function describing the spectrum of a blackbody at temperature T. Equation (23) may now be subjected to the same interpretation as already given and substituted into Eq. (20) in place of $N_1(\lambda)\ \exp[-k(\lambda)cL]$ to give the response of the instrument in the various modes of passive operation. Thus for an upward looking sensor at $\lambda > 4\ \mu m$ the emission term of Eq. (23) dominates and $N_1(\lambda,T_s) = \varepsilon_g B(\lambda,T_g)$ is substituted in Eq. (20). When a cool gas is viewed against a warm background the absorption term is important and $N_1(\lambda,T_s) = \varepsilon_b(\lambda)B(\lambda,T_b)\ \exp[-k(\lambda)cL]$. Although these two simple cases are often applicable, it is clear that in general emission and absorption occur concurrently, and that scattering and spatial temperature variations must be taken into account in any data interpretation.

B. Cell Optimization

In designing the gas filter correlation spectrometer, considerable attention must be given to the sample cell. Its transmittance depends upon pressure, temperature, and cell length. These must be controlled so as to increase S in Eq. (21) and hence S/N. The optimization procedure described here is taken from Ludwig (1972).

The transmittance of the cell in frequency units ν is given by

$$\tau_1(\nu) = \exp[k(\nu)cLp_t]$$

where p_t is the total pressure inside the cell (the sum of the partial pressures of target gas and buffer gas). The absorption coefficient $k(\nu)$ is computed for a rotational line of the gas from the expression

$$k(\nu) = \frac{S\alpha/\pi}{\alpha^2 + (\nu - \nu_o)^2}$$

where

S is the line strength $(\text{atm}^{-1} \cdot \text{cm}^{-2})$

α is half the line width at half the line strength (cm^{-1})

ν_o is frequency at the line center (cm^{-1})

The line half-width α includes pressure and temperature broadening effects. The pressure dependence is expressed in terms of an equivalent pressure

$$p_e = p_g K + p_b$$

where p_b and p_g are the partial pressures of the buffer and target gases, respectively, and K is the experimentally determined ratio of self-broadening efficiency to buffer gas broadening efficiency. Then

$$\alpha = \alpha_o p_e \frac{T_o}{T}$$

where α_o is the line half-width at STP conditions, i.e., T_o = 273.16 K. The cell transmittance can then be computed for any values of p_t, c, L, and T.

Good correlation and hence good S/N results from matching the instrumental correlator to the incoming spectrum as nearly as possible.

When the sample cell conditions are chosen it is essential to consider the design of the filter which selects the portion of the target spectrum that is to be used. Realistic values of $\beta(\lambda)$ can be selected by simple inspection of target gas and interferent gas spectra or detailed numerical calculation.

C. Instrumentation

1. Active Mode Analyzers

A number of gas filter correlation spectrometers have been built for infrared active mode gas detection. An extensive bibliography is given by Wright (1972). The two basic types employ selective cells (negative filters) or selective detectors (positive filters). A device of the former kind was first described by Schmick in 1926 (Vasko, 1963). The BRL GASPEC, also of the former kind, has been used primarily in passive sensing and is described in detail in the next section. The latter kind, the positive filter analyzer, is best exemplified by the original instrument (1938) of Luft (Wright, 1972). A diagram of the Luft analyzer is shown in Fig. 15. The detector is a differential gas thermometer consisting of two

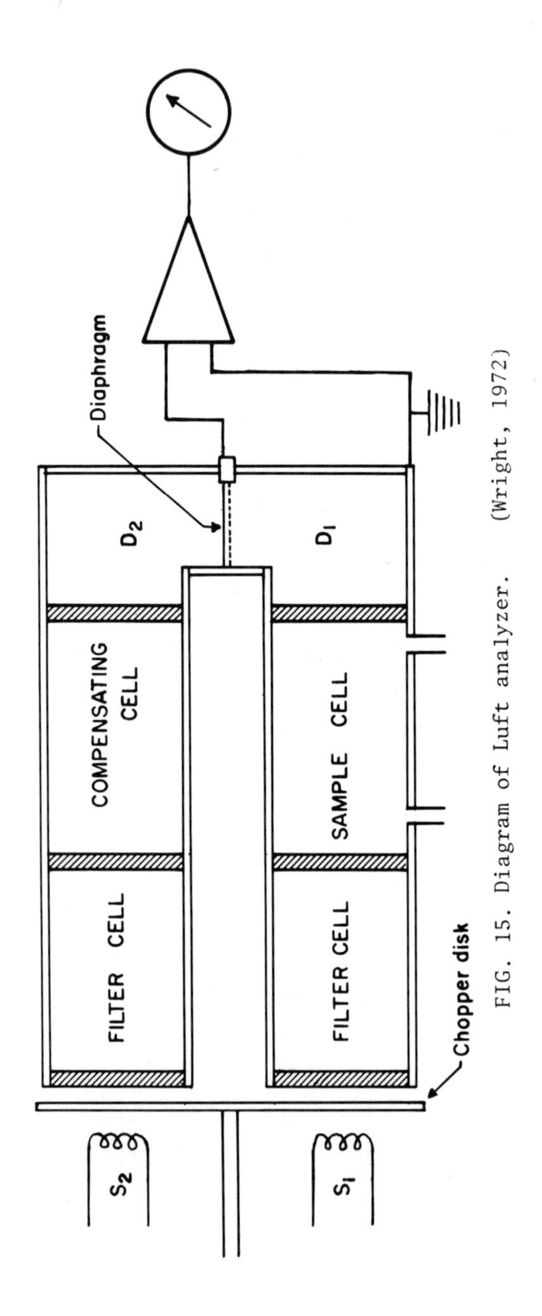

FIG. 15. Diagram of Luft analyzer. (Wright, 1972)

transparent chambers. The two chambers are separated by a flexible metal diaphragm and filled with a buffer gas containing an optimum concentration of target gas. The target gas selectively absorbs radiation entering it at its own spectral wavelengths and converts the absorbed radiant energy to thermal energy, causing the gas to expand. In front of the two detector chambers are placed a reference cell and a test cell. When contaminated air, that is, air containing a small amount of target gas, is drawn into the test cell the light transmitted through it contains fewer photons to heat the detector than does the light transmitted through the reference cell of uncontaminated air. The gas in chamber 1 thus undergoes a greater increase of pressure than that in chamber 2 and the diaphragm moves. If now a rigid wire gauze is set in chamber 2 parallel to but not in contact with the diaphragm, one can simply measure the capacitance between the gauze and the diaphragm to detect changes of target gas concentration in the single cell.

2. Passive Mode Sensors; the BRL GASPEC

Gas filter correlation spectrometers of the negative filter type have been designed and built under the trade name GASPEC by Barringer Research for the passive remote detection of trace vapors. The present design is shown in Fig. 16. The GASPEC employs two separate detectors, D_1 behind the optimized sample cell and D_2 behind the spectrally neutral reference cell. A chopper disc assures the fluctuating radiance needed for efficient operation of the infrared detectors. Light from the source via lens L1 and the chopper disc is separated into two paths toward the two detectors by a beamsplitter. With these components two alternating signals are produced at the two detectors, the amplitude difference of which obeys Eq. (20).

The light from the source is chopped at one frequency while the reference sources are chopped at a different frequency. This reference signal is superposed upon both source signals from detectors D_1 and D_2. The source and reference signals V + R (following the block diagram of Fig. 17 from the detectors are subtracted and passed to two separate product detectors. The product detectors separate the

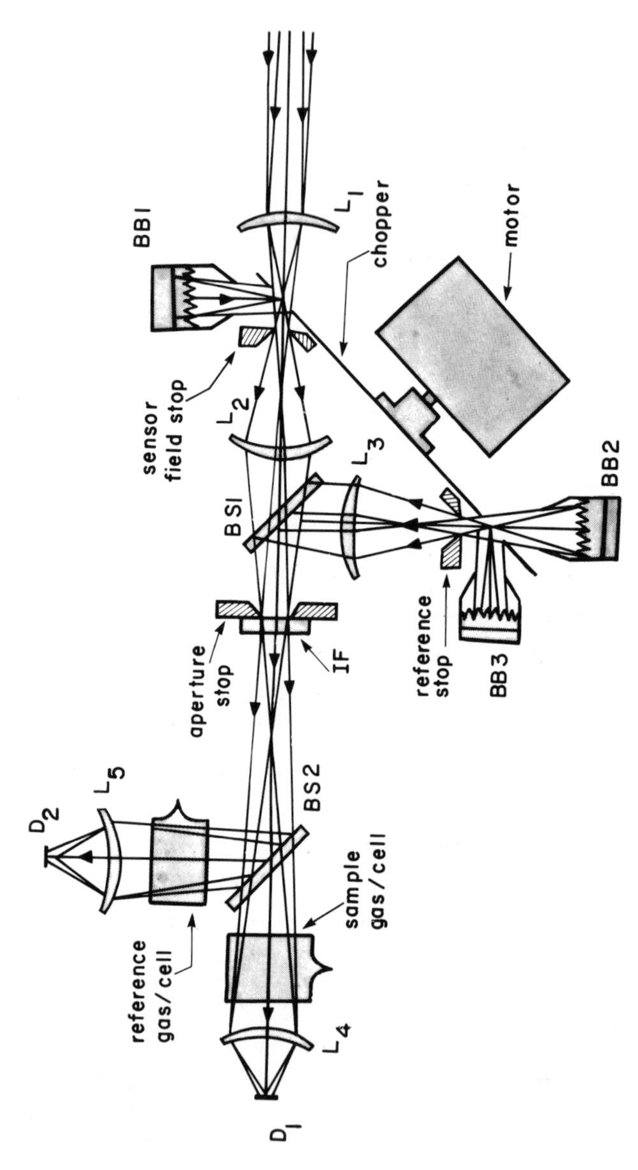

FIG. 16. Optical layout of GASPEC.

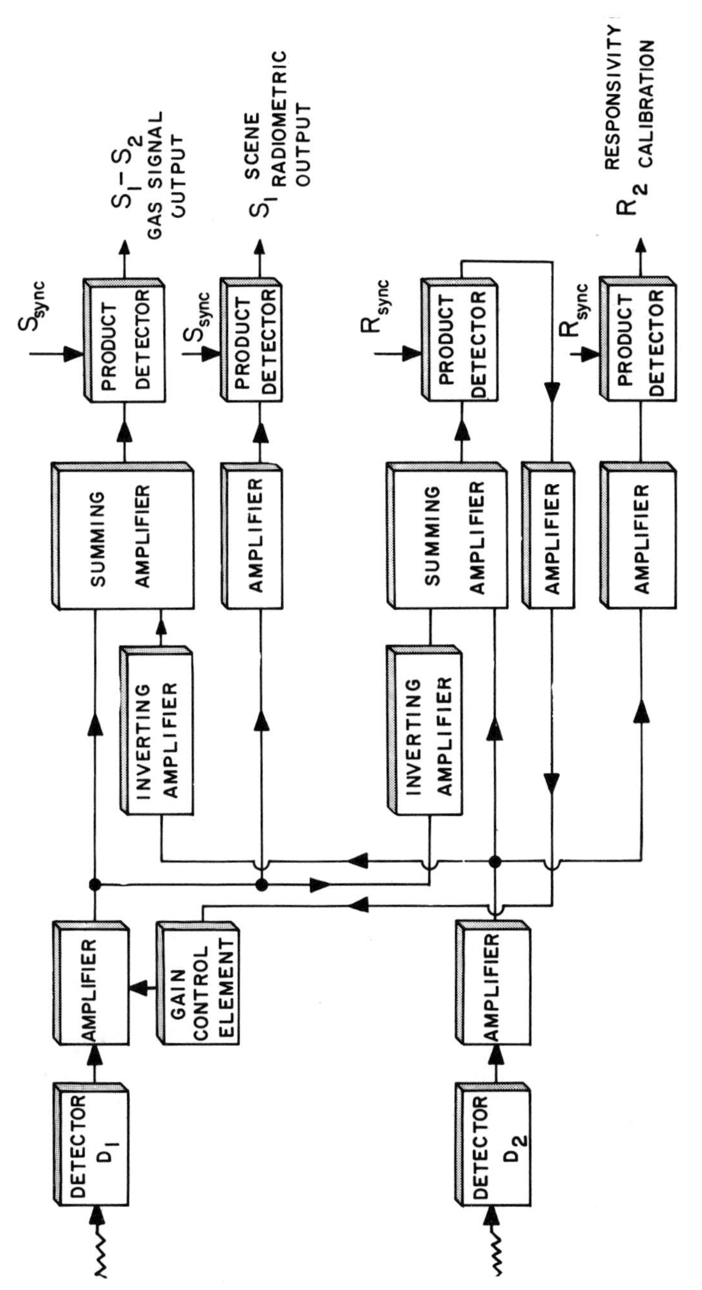

FIG. 17. Electronic signal processing block diagram for GASPEC.

high and low frequency components of the signals so that one channel is left with the signal difference $V_1 - V_2$ and the other with the reference difference, $R_1 - R_2$. The source signal difference is passed through an integrator and DC amplifier and finally to a chart recorder. The reference difference is compared in the error amplifier with gain trim adjust control. The error amplifier drives the AGC in the first channel to bring the error signal to zero. In this way the two channels are always balanced, and the cell obstruction problems of earlier negative filter devices (Wright, 1972) are avoided. The gain trim adjustment is made while pointing the instrument through an atmosphere without target gas and sets the response equal to zero as described in Sec. III.B.

Two blackbody radiators BB1 and BB3 are included in the system as shown in Fig. 17. They provide known, stable radiances when the source and BB2 are hidden by the chopper. BB1 performs the additional service of keeping the modulated source voltages at a convenient level for differential detection of the very small gas signals.

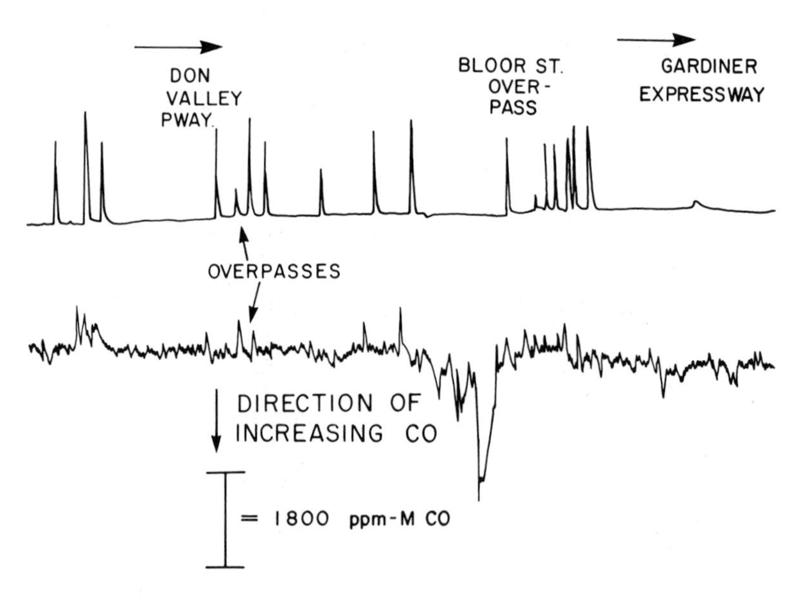

FIG. 18. Truck mounted CO measurements. (Ward and Zwick, 1975)

A sample record from a GASPEC designed to detect CO is shown in Fig. 18 (Ward and Zwick, 1975). It was taken while the instrument was carried looking upward in a truck throughout the metropolitan Toronto area. The same authors report that a similar instrument was successfully used in the detection of leaks in natural gas lines (CH_4) in Alberta.

IV. CORRELATION INTERFEROMETRY

A. General

Fourier spectroscopy or interferometry has during the past two decades grown into a powerful and widely used technique for chemical analysis. Its complete multiplexing of all spectral channels and its large throughput (the Felgett and Jacquinot advantages, respectively) give it a well-documented superiority over the conventional scanning spectrometer for weak sources of light. A Fourier spectrometer is in a sense a correlation spectrometer in that the interferogram it produces is the autocorrelation function of the spectrum. That is, instead of correlating the incoming spectrum with a replica spectrum, it correlates the incoming spectrum with itself. The correlation interferometer carries this a step further by correlating the incoming interferogram with a replica interferogram stored in the instrument.

The correlation interferometer described here is first of all a Michelson interferometer, drawn schematically in Fig. 19. The interferogram is produced by varying the optical path of one of the arms of the interferometer. Every change of arm length by one-half the wavelength of the incoming radiation produces a complete fringe scan at the detector and therefore a complete cycle of electrical oscillation as the recombined beams from the two arms reinforce, cancel, and then reinforce again. A simple emission spectrum consisting of a single monochromatic line gives an interferogram signal which is a cosine wave train of constant amplitude and frequency $u\nu_o$, where u is the speed with which the path length is changed (delay scanning velocity) and ν_o is the optical frequency (cm^{-1}) of the line.

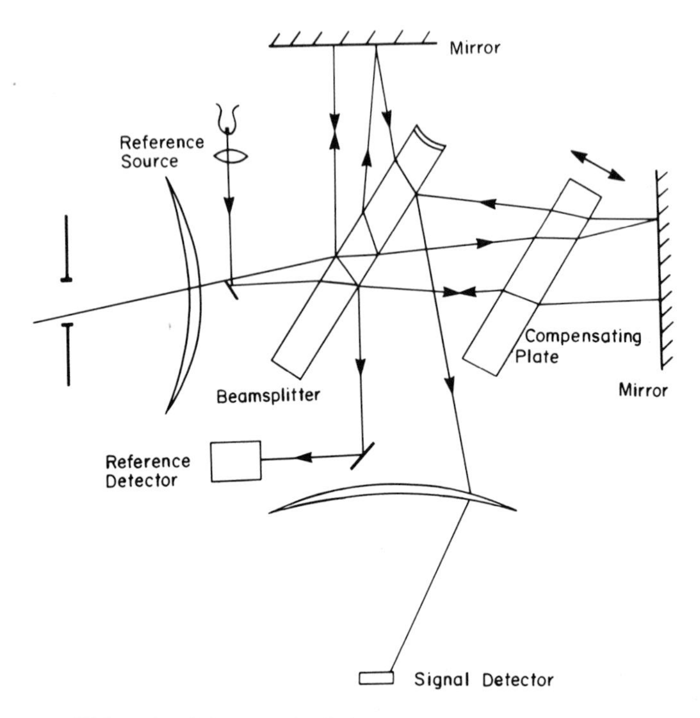

FIG. 19. Scheme of Michelson interferometer.

This waveform is shown in Fig. 20. If the spectrum consists of two closely spaced lines of equal intensity, the same type of cosine pattern develops for either but both have their own characteristic frequencies, $u\nu_0$. The detector receives the vector sum of the two wave trains which is the interferogram shown in the middle of Fig. 20. It can be characterized by a cosine wave of frequency equal to the average frequency of the two constituent wave trains and amplitude that varies at a frequency which is the difference between those of the component wave trains.

The relationship between the spectrum $N(\nu)$ and its interferogram $I(X)$ is usually expressed by saying they are Fourier transforms of each other. That is

$$I(X) = \int_0^\infty N(\nu) \cos(2\pi\nu X)\, d\nu$$

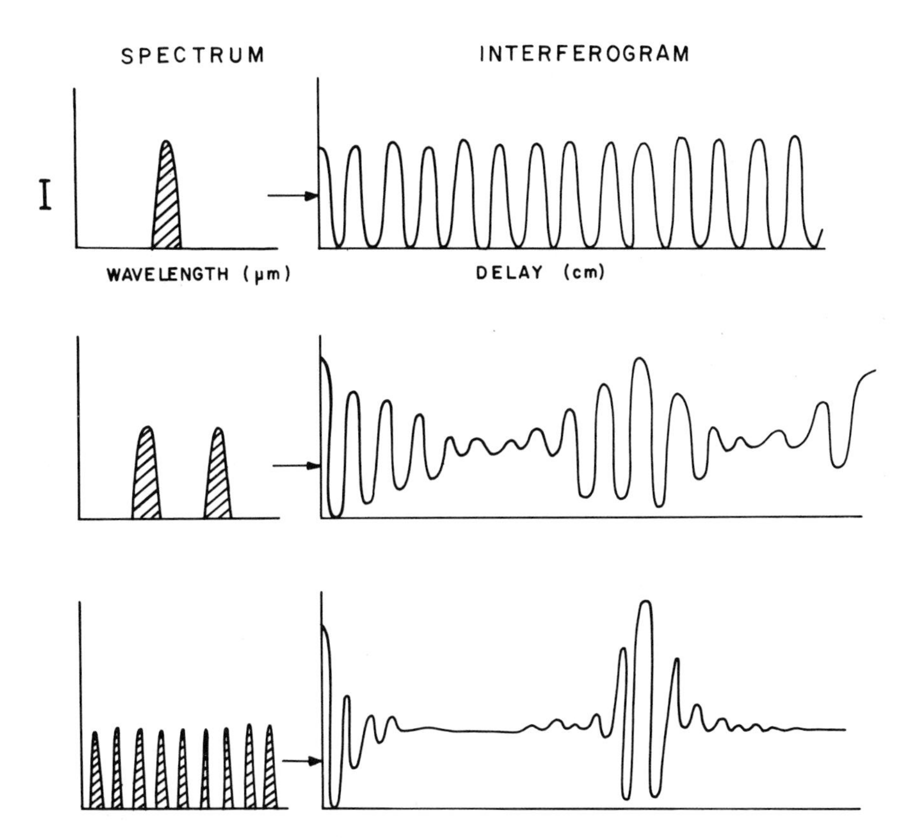

FIG. 20. Waveform of correlation interferometer.

is the intensity of the recombined light beam when the optical path difference between the two arms or the delay is X. N(λ) here represents the spectral radiant power available to the detector.

The spectrum shown at the bottom of Fig. 20 is a Dirac comb, an array of equally spaced lines (constant frequency interval) of equal amplitude. The Fourier transform of an infinite Dirac comb is another Dirac comb in delay space, the spacing between spikes being $1/\Delta\nu$, or the reciprocal of the spectral spacing. If the Dirac comb is not infinite but is limited in extent, the spectrum is best represented as the product of the Dirac comb and a rectangular function. The Fourier transform of the product is the convolution of the delay Dirac comb and the sin X/X function (the Fourier transform of the rectangular function). The resulting interferogram is a series of equally spaced sin X/X functions as shown.

Carbon monoxide has received the most attention with regard to the correlation interferometer. Its spectrum at 2.3 μm, shown in Fig. 21, consists of a series of almost equidistant lines, i.e., it is approximately a Dirac comb bounded at both ends by a filter. Its interferogram, also shown in Fig. 21, therefore approximates that of the Dirac comb in Fig. 20 with envelope maxima at 0, $2\frac{1}{4}$, and $4\frac{1}{2}$ mm. The spacing is a result of the fairly regular spacing of the spectral lines which is on the average 4.4 cm^{-1}. The fact that the amplitude and spacing of the spectrum are not regular can lead to some reduction of the interferogram maxima. Either the spectrum or the interferogram can be used to identify CO, for the same information is contained in both. If the interferogram is used, the intensity at any maximum of the envelope can be measured and considered proportional (for small amounts of CO) to the CO burden.

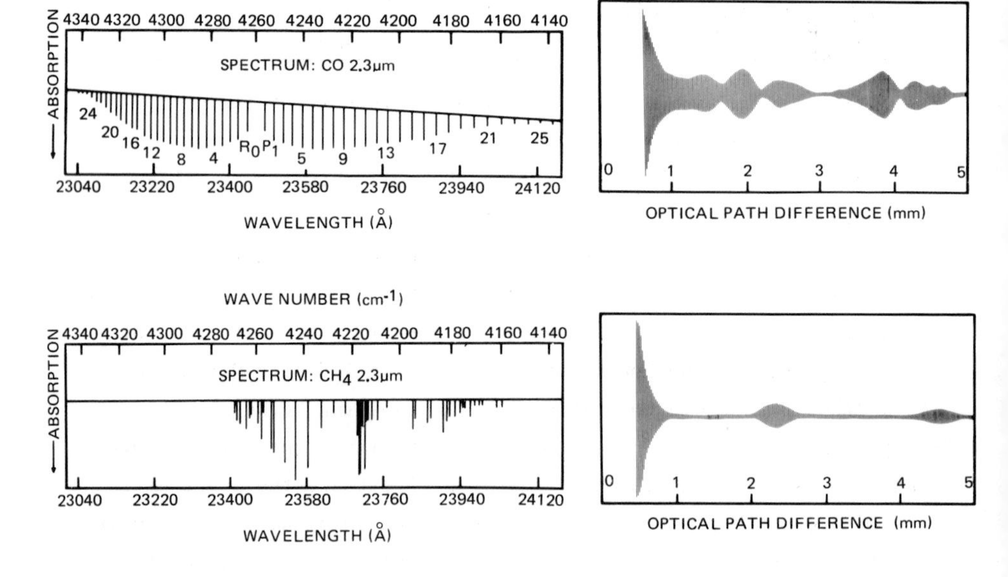

FIG. 21. Spectra and interferograms of carbon monoxide and methane. (Bortner et al., 1973)

Thus a fixed interferometer could be built with the delay set at a maximum of the interferogram. However, any interfering spectra can also affect this maximum so that the fixed point interferometer is not practical. In the case of CO in the atmosphere the main interferents are CH_4 and H_2O, both of which are generally more abundant than the CO target gas.

The spectrum of CH_4 is shown in Fig. 21 for the same spectral region as its interferogram. To resolve the CO and the methane by means of the spectra shown, a spectral resolution of 0.2 cm^{-1} would be required. This could be achieved by means of a high-resolution spectrometer or by collecting the interferogram over a delay range from 0 to 5 cm. However, since most of the CO contribution to a combined interferogram is contained in the small delay regions of the maxima at 2.225 and 4.50 mm, there is no need for the delay scan to include any more than these regions. In fact scanning between 2.0 and 3.0 mm provides all the information necessary to distinguish between CO and CH_4. In the correlation interferometer such a limited delay range is scanned and the shape of the interferogram envelope over this limited range is compared with a correlator function somewhat like the corresponding interferogram region of pure CO. This technique thus offers all the advantages of standard Fourier spectroscopy for low light levels, but avoids the difficulty of requiring computation of a long interferogram to transform it to a spectrum. Moreover, the short scan requires less sophistication mechanically than does the long scan and leads to a more stable instrument than the conventional Michelson interferometer.

B. Analysis

The spectral radiant power available to the detector at any delay X is given by the equation

$$I(X) = A\Omega \int_0^\infty N(\nu)\beta(\nu)e^{-k(\nu)cL}[1 + \cos(2\pi\nu X)]\, d\nu \tag{24}$$

The interferogram thus includes the Bouguer-Beer law dependence on cL. The effect of interfering gases can be included in the spectral source function $N(\nu)$ or it can be stated explicitly by multiplying the transmittance of each interferent, $\exp[-k_i(\nu)c_iL]$, by $N(\nu)$. The final signal is produced by multiplying the interferogram by some correlator function $H(X)$ and integrating the product over the whole scan.

$$M = \int_{X_1}^{X_2} H(X)I(X)\, dX \qquad (25)$$

If average absorption coefficients are used for the target gas as was done for the dispersive and nondispersive correlation spectrometers, the signal is again exponentially dependent upon cL.

Detector noise should be the limiting noise mechanism in any measurement by the infrared correlation interferometer. It is described by the same equation as that used in Sec. III, i.e.,

$$NEP = \frac{\sqrt{A_d \Delta f}}{D_\lambda^*}$$

Again A_d is the detector area, Δf is the bandwidth of the processing electronics and D_λ^* is the detectivity of the detector. A number of other noise sources are also possible, however, and although effort is made to minimize them it has been found convenient to compensate them in the correlator function. They are best treated by grouping them into four categories. The a_i are severity factors for each component.

a_1 Random Additive (RA)
- detector noise
- photon noise
- electrical noise
- digitizing of interferogram

a_2 Random Multiplicative (RM)
- scintillation of incident light
- reference signal errors
- scan waveform variations

a_3 Systematic Additive (SA)
- spectral elements not included in correlator determination
- cross-talk from reference into signal channel

a_4 Systematic Multiplicative (SM)
- computer truncation of correlator components
- scan function different from that used in correlator determination

The systematic errors are a type of synchronous noise, for which $\Delta N(X)$ is directly proportional to the time of observation $D(X)$. The random noise contributions are proportional to $\sqrt{D(X)}$. The square of the total noise at a point, $\Delta N^2(X)$, in the interferogram with intensity $\langle I(X)\rangle$ is the sum of the squares of these individual noise contributions:

$$\Delta N^2(X) = RA^2 + RM^2 + SA^2 + SM^2$$

$$\Delta N^2(X) = a_1 H^2(X)D(X) + a_2 H^2(X)D(X) \overset{X}{<} + a_3 H^2(X)D^2(X)$$
$$+ a_4 H^2(X)D^2(X) \langle I(X)\rangle^2$$

or

$$\Delta N^2(X) = H^2(X)[\{a_1 + a_2 \langle I(X)\rangle^2\}D(X) + \{a_3 + a_4 \langle I(X)\rangle^2\}D^2(X)]$$

Substituting

$$G_1(X) = a_1 + a_2 \langle I(X)\rangle^2$$
$$G_1(X) = a_3 + a_4 \langle I(X)\rangle^2$$

gives

$$\Delta N^2(X) = H^2(X)D(X)[G_1(X) + G_2(X)D(X)]$$

Finally, the total noise in a measurement of target gas concentration is

$$N = \{\int H^2(X)D(X)\ G_1(X) + G_2(X)D(X)\}^{1/2}\ dX \tag{26}$$

The quantities $G_1(X)$ and $G_2(X)$ now incorporate all types of error and serve as noise equivalent interferograms.

C. Determination of the Correlator Function

It is clear from the description of the instrument that the central feature of the correlation technique is the correlator function. In principle this function need only be a reproduction of a pure target gas interferogram. In practice the correlator function is flexible enough that the effects of spectrally interfering gases and many kinds of noise can be negated. The following analysis is that of Levy (Bortner et al., 1973).

The interferogram of natural light may be regarded as the sum of some "nominal" interferogram $I_0(X)$ and a number n of "disturbed" interferograms linearly related to the nominal. That is the interferogram is represented by its first-order Taylor expansion.

$$I(X) = I_0(X) + \sum_{i=1}^{N} q_i I_i(X)$$

The response M of the interferometer can then be written [including now the effect of the duration of the measurement D(X)]

$$M = \int H(X)D(X)I(X)\ d\chi = \int H(X)D(X)I_0(X)\ d\chi + \int \sum_{i=1}^{N} q_i H(X)D(X)I_i(X)\ d\chi$$

or more briefly

$$M = M_o + \Sigma q_i M_i$$

Eliminating the effects of everything except the target gas, which we denote by i = 1, means that we impose the conditions

$$M_i = 0 \qquad i \neq 1$$
$$M_i = 1 \qquad i = 1$$

The response is then just

$$M = M_o + q_1$$

where q_1 will be proportional to the optical depth of target gas. We wish to satisfy this condition while at the same time keeping the noise to a minimum. Much of this can be done by choosing the D(X) carefully, for which a procedure is described by Bortner et al. (1973). However, the analysis is much simpler if D(X) is given as a specified constant, and since this alternative was chosen in all instruments designed so far, it will be followed here. The mathematical problem thus can be stated

$$\text{Minimize } N^2 = \int H^2(X)G(X)\ d\chi \qquad (27)$$

with the condition that

$$\int H(X) I_i(X) dX = \delta_{i1} \quad (28)$$

Here the noise, Eq. (26), has been rewritten with G(X) incorporating the more specific $G_1(X)$ and $G_2(X)$. The Kronecker delta, δ_{i1}, has its usual meaning.

Equation (28) is multiplied by the Lagrange multiplier m_1 and added to Eq. (27).

$$\int H^2(X) G(X) dX + \Sigma\, m_i \int H(X) I_i(X) dX = \text{minimum}$$

Differentiating with respect to each "weight" $H(X_k)$ and equating to zero gives

$$\frac{\partial}{\partial H(X_k)} \left(\int H^2 G\, dX + \Sigma\, m_i \int H I_i\, dX \right) = 2H(X_k) G(X_k) + \Sigma\, m_i I_i(X_k) = 0$$

or

$$H(X_k) = \left[\sum \frac{m_i}{2} I_i(X_k) \right] / G(X_k)$$

These values of $H(X_k)$ may now be substituted in Eq. (28) to solve for m_i.

$$\int H(X) I_j(X)\, dX = \delta_{j1} = \sum \frac{m_i}{2} \int \frac{I_i I_j}{G}\, dX$$

Defining

$$A_{ij} = \int \frac{I_i I_j}{G}\, dX \quad (29)$$

this becomes

$$\sum \frac{m_i}{2} A_{ij} = \delta_{ji}$$

which is a set of linear equations soluble by matrix inversion

$$m_i = 2 \sum_{j=1}^{n} A_{ij}^{-1} \delta_{ji}$$

$$H(X) = 2 \sum_{i=1}^{n} A_{i1}^{-1} \frac{I_i(X)}{G(X)} \quad (30)$$

This method of computing the correlator function H(X) requires a knowledge of a range of interferograms and noise levels as basic input, i.e., $I_i(X)$ and $G(X)$ must be known. The $I_i(X)$ were found by measurements on mixtures of CO, H_2O, and CH_4 in the following manner:

1. A "nominal" test condition was set up in a laboratory gas cell containing 3 atm-cm of methane and 0.2 atm-cm of carbon monoxide which is equivalent to what a satellite-borne instrument would see in looking down through a nominal atmosphere. An average interferogram was obtained from the test cell using a glowing filament light source.
2. A "target" test condition was then set up in the same gas cell with the same concentration of CH_4 but with 0.3 atm-cm of CO. Another average interferogram was taken and stored.
3. A series of test conditions produced a corresponding series of interferograms as the CH_4 optical depth was varied from 1 to 5 atm-cm with the CO held at 0.2 atm-cm.
4. A series of tests was run outside on the real atmosphere with and without a cell of CO in the light path, and average interferograms were again stored.
5. The nominal interferogram was subtracted from each of the test and target interferograms to provide the difference interferograms $I_i(X)$.
6. Realistic noise values were assumed for a_1, a_2, a_3, and a_4.
7. The matrix A_{ij} was computed for all possible combinations of test conditions and noise according to Eq. (29). This matrix was inverted and used in Eq. (30) to compute $H(X)$.

The correlator function determined in this way is multiplied by any incoming interferogram and integrated over the useful delay range giving $M = 0$ for the "nominal" test condition and $M = 1$ for the "target" test condition.

D. Instrumentation

1. Optical

A progressive series of correlation interferometers have been built by Barringer Research primarily for participation in a GE-NASA experiment to locate sources and sinks of CO about the earth from a satellite. One such instrument is that described by Bortner et al. (1973) and shown in block diagram form in Fig. 22.

Incident radiation is collected by a 200 mm, f/2 Newtonian telescope with a 2° field stop. A temperature-controlled interference

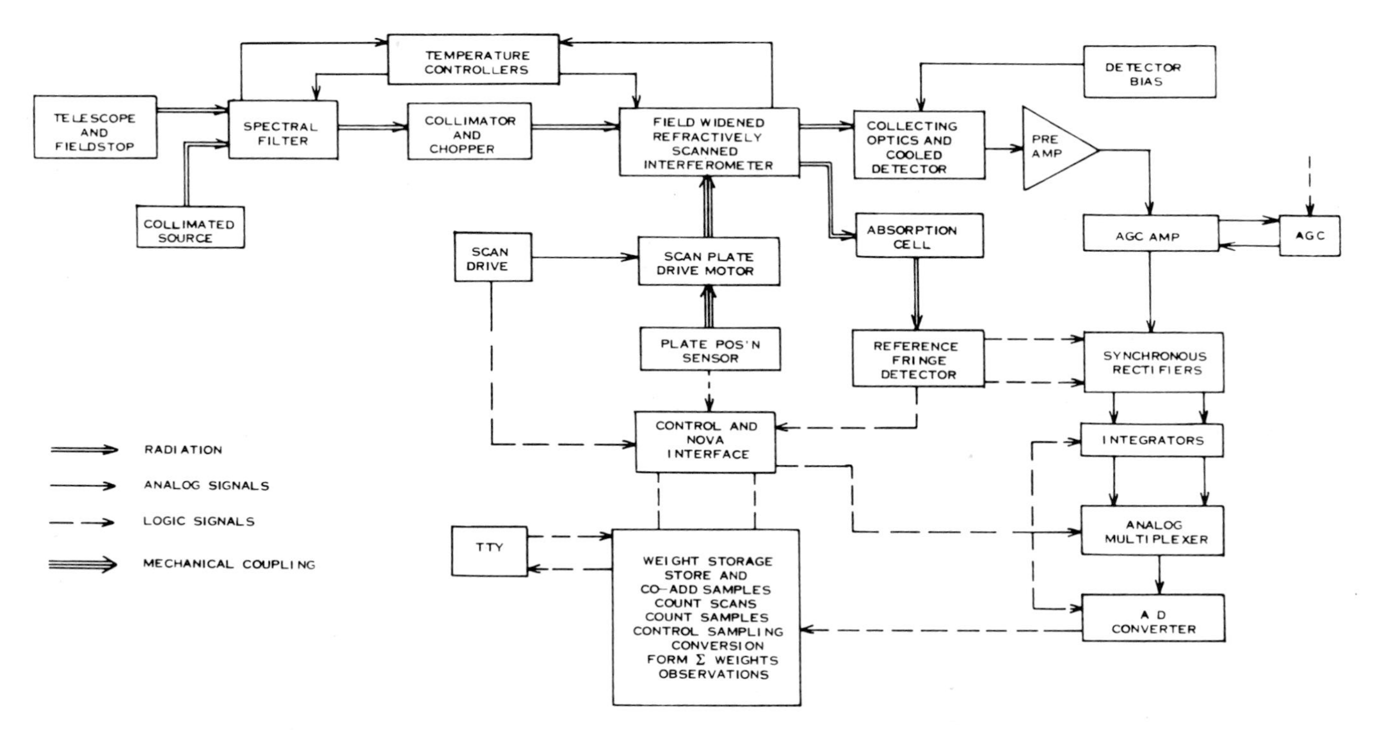

FIG. 22. Block diagram of Barringer correlation interferometer for CO detection. (Bortner et al., 1973)

filter centered at 4280 cm^{-1} with a width at half-transmission of 45 cm^{-1} is placed just in front of the focal plane of the telescope. A chopper wheel rotates in the focal plane of the telescope, driven by a stepping motor. The light is "collimated" by a silicon lens which actually spreads the beam to 7° making full use of the field-widened design.

Field widening, described by Hilliard and Shepherd (1966) and by Steel (1967), is accomplished by the use of a thick mirror in one of the interferometer arms. The optimization of the thick mirror is described in detail by Bortner et al. (1973). The beam splitter used in the interferometer is silicon coated with aluminum.

The short delay scan required by the correlation technique is accomplished by rotation of the compensating plate. The compensating plate is normally introduced into one of the arms to make the geometric paths equal when the optical paths are equal and is therefore equal in thickness, composition, and incidence angle to the beam-splitter block. The advantage of scanning by plate rotation over scanning by mirror translation as is usually done is that accurate rotation is easier than accurate translation. The refractive scanning requires a large rotation to produce a small change of delay, and since the compensator plate is the only moving part of the instrument, a mechanically stable interferometer is possible. The scanning plate is driven by a torque motor governed by an exponential waveform that causes a linear decrease of delay with time for 3/4 the scan period. The remaining 1/4 of the scan is used for rapid return to the zero position.

After the two beams are recombined at the beamsplitter the fringe pattern is focussed by another silicon lens onto a lead sulfide detector immersed in liquid nitrogen. Care has been taken to minimize the area of the detector while maintaining maximum throughput in order to attain the best possible signal-to-noise ratio.

It is important in any interferometer that no relative motion of optical components take place except those expressly intended. This means that the interferometer be made as insensitive as possible to mechanical shock and thermal expansion. The present instrument is made of fused titanium silicate (Corning ULE) which is mechanically stiff and has a thermal expansion coefficient of only 3×10^{-8} deg^{-1}. (That of Invar is 1.5×10^{-6} deg^{-1}.)

An independent reference channel coaxial with the main optical channel is included in the system for signal processing purposes. The reference light beam originates at an incandescent lamp located in the center of the telescope primary and passes through the whole interferometer along with the main beam. Upon leaving the beam-splitter the recombined reference beams pass through a gas cell filled with methane to a separate lead sulfide detector at room temperature. The signal received by this detector is the interferogram of methane with each fringe exactly in phase with the corresponding fringe of the main interferogram regardless of any irregularity in the scan form. One maximum occurs in the methane interferogram over the delay region scanned and it is used as the absolute reference for the scan drive system. Delay is measured from this maximum by an electronic fringe counter.

2. Electronic

The electronic block diagram for the correlation interferometer is incorporated into Fig. 22. The task of the electronic system is to sample the wave train, that is, the interferogram and record it in digital form. However, the high-frequency interferogram contains much that is redundant in that the individual oscillations are of constant frequency characteristic only of the central wavelength of the optical filter. The important information is contained in the envelope of the wave train. The situation is similar to the reception of amplitude modulated radio waves and the demodulation technique may therefore be the same. The reference interferogram is used to generate two wave trains, one exactly in phase with the main interferogram and a second leading the first by $\pi/2$. These correspond to the in-phase and quadrature-phase reference signals in a homodyne detector. They are both multiplied by the main interferogram in the synchronous rectifiers and integrated before being added together in the analog multiplexer where they form a wave train with the shape of the envelope of the main interferogram. This wave is sampled by the analog-digital converter and transferred to a NOVA minicomputer. The reason for including the quadrature phase reference signal is to give a positive signal-to-noise ratio even when the in-phase signal goes to zero. An automatic gain control network is provided in the signal channel to allow the instrument to work properly over a wide range of source brightness, such as may be encountered in a satellite-borne instrument. While the scanning plate is in the first (return) quarter of its cycle, the detector output is integrated and the result is compared with a reference dc level in an error amplifier. The difference is used to adjust the voltage across the detector until zero difference is attained. The adjusted detector voltage is then held constant during the forward scan of the compensator plate.

The portion of interferogram sampled is controlled by a "run" pulse initiated when the scan plate passes a position that permits a light emitting diode to shine on a photodiode sensor. The run pulse stops when the number of reference fringes corresponding to the desired delay change are counted from the beginning of the "run" pulse. The peak in the envelope of the reference signal is used as a reference giving the delay position ±3 fringes. As each fringe is detected its amplitude is compared with that of the preceding fringe and held in a sample and hold circuit. When a local maximum has been detected, a pulse is generated which can be compared with the start of the "run" pulse.

In this particular model of the correlation interferometer the digitized interferogram envelope is read by an on-line NOVA minicomputer, which then performs the multiplication by the similarly digitized correlator function and the final integration and prints out the result, M. This was a particularly useful approach when a generally applicable correlator function had not yet been devised and a program for including new observations in the correlator was necessary. A more recent model simply stores the digitized interferograms on magnetic tape for subsequent processing on a large computer.

Without too much difficulty a computation network can be built into the instrument, thus eliminating the need for an outside computer. Circuits for this type of processing are discussed by Bortner et al. (1973).

3. Application

An instrument of the type just described was built and tested in a cooperative program between BRL, General Electric, and NASA and the results were described by Goldstein et al. (1973). Flying the interferometer over eastern Canada in an aircraft at 28,000 ft in the downward-looking mode, they found CO concentrations between 48 and 315 ppb depending upon the proximity of large urban areas. The concentration of CO was more variable than that of CH_4 which covered a range from 1.09 to 2.19 ppm. Measurements of NO, SO_2, NH_3, NO_2, N_2O, CO_2, and H_2O are also regarded as feasible targets by Goldstein et al. (1973).

V. CONCLUSIONS

It is seen from the preceding chapters that correlation spectroscopy provides a variety of methods for the remote detection of trace gases in the atmosphere and that these methods are conveniently effected by three basic types of instruments. At BRL these three instruments, all in different stages of development, continually are reviewed and improved. It is therefore inappropriate in a study of this sort to describe in detail the performance specifications of each, which can be found in periodicals and company literature (e.g., Wiens et al., 1975; Ward and Zwick, 1975). Some qualitative intercomparison of the three instruments could be helpful, however.

The correlation interferometer and the gas cell correlation spectrometer both operate well in the infrared with approximately the same throughput and sensitivity for equal resolution. Because the gas cell instrument is considerably simpler than the interferometer it is correspondingly less expensive and therefore in most cases preferred. However, a correlation interferometer provides an unequivocal separation of target and interferent gas signatures and would certainly be preferable when interference renders the data from a gas filter spectrometer ambiguous. The correlation spectrometer is generally not useful in the infrared because at the high resolutions necessary to avoid interference the throughput is significantly lower than for the other two instruments. For gases whose spectra exhibit so much overlapping of adjacent lines that high resolution measurements provide no real benefits (e.g., SO_2 in the ultraviolet) the rugged and portable correlation spectrometer is the best choice. The final choice of instrument can only be made, of course, by detailed comparison of the sensor characteristics and the circumstances of the measurement.

REFERENCES

A. R. Barringer and J. P. Schock, Progress in the remote sensing of vapours for air pollution, geologic, and oceanographic applications, *Proceedings of the Symposium on Remote Sensing of the Environment*, University of Michigan, Ann Arbor, pp. 779-792, 1966.

Ely E. Bell, Leonard Eisner, James Young, and Robert Oetjen, Spectral radiance of sky and terrain at wavelengths between 1 and 20 microns. II. Sky measurements. *J. Opt. Soc. Am., 50*, 1313-1320 (1960).

M. H. Bortner, R. Dick, H. W. Goldstein, R. N. Grenda, and G. M. Levy, Development of a breadboard model correlation interferometer for the carbon monoxide pollution experiment. General Electric NASA Report CR-112212, 1973.

M. Bottema, J. Plummer, and J. Strong, *Astrophys. J., 139*, 1021 (1964).

Norman Braslau and J. V. Dave, Effect of aerosols on the transfer of solar energy through realistic model atmospheres. Part III Ground level fluxes in the biologically active bands .2850-.3700 microns. IBM Report RC 4308, 1973.

Kurt Bullrich, Scattered radiation in the atmosphere and the natural aerosol. *Advances in Geophysics*, Vol. 10 (H. E. Landsberg and J. Van Miegham, eds.), Academic Press, New York, pp. 99-260, 1964.

R. Dick and G. Levy, Correlation interferometry. Presentation at the International Conference on Fourier Spectroscopy, Aspen, Colorado, 1970.

M. Harwit and J. A. Decker, Modulation techniques in spectrometry. *Progress in Optics*, Vol. 12 (E. Wolf, ed.), North-Holland, Amsterdam, 1975.

H. W. Goldstein, M. H. Bortner, R. N. Grenda, A. M. Karger, R. Dick, F. David, P. J. LeBel, The remote measurement of trace atmospheric species by correlation interferometry. Presented at Second Joint Conference on Remote Sensing of Environmental Pollution, Washington, 1973.

R. M. Goody, *Atmospheric Radiation I: Theoretical Basis*, Oxford University Press, London, 1964.

Richard Goody, Cross-correlating spectrometer. *J. Opt. Soc. Am.*, *58*, 900-908 (1968).

Robert N. Hager, Jr., Derivative spectroscopy with emphasis on trace gas analysis. *Anal. Chem.*, *45*, 1131A-1138A, 1973.

A. W. Harrison and D. J. W. Kendall, Fraunhofer line filling-in (3855-4455 Å). *Can. J. Phys.*, *52*, 940-944 (1974).

R. L. Hilliard and G. G. Shepherd, *J. Opt. Soc. Amer.*, *56*, 362 (1966).

J. A. Hodgeson, W. A. McClenny, and P. L. Hanst, Air pollution monitoring by advanced spectroscopic techniques, *Science*, *182*, 248-258, 1973.

R. B. Kay, Absorption spectra apparatus using optical correlation for the detection of trace amounts of SO_2, *Appl. Optics*, *6*, 776-778 (1967).

K. Ya. Kondratyev, *Radiation in the Atmosphere*, Academic Press, New York, 1969.

C. B. Ludwig, Monitoring of air pollution by satellites (MAPS). Science Applications, NASA Report NAS1-12048, 1973.

Millán M. Millán, A study of the operational characteristics and optimization procedures of dispersive correlation spectrometers for the detection of trace gases in the atmosphere. Thesis, Univ. of Toronto, 1972.

Millán Muñoz Millán and Gilbert S. Newcomb, Theory, applications, and results of the long-line correlation spectrometer, *IEEE Trans. Geosci. Electron.*, *GE-8*, 149-157 (1970).

Andrew J. Moffatt and Millán M. Millán, The application of optical correlation techniques to the remote sensing of SO_2 plumes using sky light. *Atmos. Environ.*, *5*, 677-690 (1971).

J. F. Noxon and R. M. Goody, *Atmos. Oceanic Phys.*, *1*, 275 (1965).

Ralph W. Nicholls and Baxter H. Armstrong, *Emission, Absorption, and Transfer of Radiation in Heated Atmospheres*, Pergamon Press, Oxford, 1972.

W. H. Steel, *Interferometry*, Cambridge Univ. Press, London, 1967.

A. Vaškŏ, *Infrared Radiation*, CRC Press, Cleveland, 1968.

T. V. Ward and H. H. Zwick, Gas-cell correlation spectrometer: GASPEC, *Applied Optics*, *14*, 1896 (1975).

R. H. Wiens, New performance characteristics of the Barringer correlation spectrometer. Barringer Research Report, 1975.

William L. Wolfe, *Handbook of Military Infrared Technology,* Office of Naval Research, Washington, D. C., 1965.

J. C. Wright, Industrial plant analysis, in *Laboratory Methods in Infrared Spectroscopy* (R. G. J. Miller and B. C. Stace, eds.), Heydon, London, 1972.

H. H. Zwick and Millán M. Millán, Feasibility study of using ground modulation to discriminate between ground and atmospheric-scattered radiation, *Can. Aeron. Sp. J.*, *17*, 413 (1971).

Chapter 4

THE LABORATORY COMPUTER AS AN INFORMATION CHANNEL

Charles T. Foskett

Digilab, Inc.
Cambridge, Massachusetts

I. INTRODUCTION

The Last Whole Earth Catalogue [1] (Access to Tools) underscores a very important fact of this latter half of the twentieth century: digital computers are available to every man. Reference is made to How to Build a Working Digital Computer [2]. This computer has a 28-bit instruction with a 10-instruction repertoire. It is an electromechanical device of metal, wood, and various components; has much size and little speed. The book costs approximately five dollars.

For the well-funded enthusiast, IBM will offer a wide spectrum of processors whose memory size, physical size, speed, instruction set, and general computational power will exceed the "Whole Earth" version by orders and orders of magnitude.

There was a time, not so long ago, when the IBM 1620 was a small but modern computer, which made modern computation techniques available to small universities, businesses, etc., at a moderate cost. In 1972, a prominent west coast semiconductor firm began offering a small programmable semiconductor chip, any two of which are roughly the equivalent of the IBM 1620. The price of this chip is less than $100. (Of course, it does not operate by itself.) A network of drivers is required, but the gigantic incremental advance of large-scale integration (LSI) in semiconductor technology is obvious.

Computer science is much like every human endeavor; it is objective and subjective. The user (the scientist, the businessman, the academician, the student) must choose the system which best suits his needs. In the laboratory, there has been, in recent years, considerable controversy over the use of centralized versus distributed computer power. The author's preference, in view of the declining costs of hardware, is toward the latter, although there is clearly an optimum middle ground.

For experiment control, for the application wherein the computer is an integral link in the signal chain, the distributed power or localized computer is the preferred approach. For those applications involving large information retrieval situations or "number-crunching" problems, the centralized computer is more desirable.

This chapter is directed toward methods of defining the laboratory experiment in such a fashion that the choice can be intelligently made. Section II is a brief and basic introduction to some concepts in information theory which are as useful in designing experiments in chemistry and physics as they are in designing radar networks or computer processors. Section III discusses computer configurations and the properties of some of the small laboratory computers. Section IV presents an approach to organizing the laboratory software system which formerly was available only on large machines. Section V considers briefly a practical application of these concepts.

II. SIGNAL CHANNELS AND INFORMATION IN LABORATORY ENVIRONMENTS

A. Message Transmission

The information content of a laboratory experiment may be concisely expressed in terms of the bandwidth and the signal-to-noise ratio of the experiment. This definition has a much more restricted meaning than the term "information" in normal English usage. The origins of this definition may be traced to Gibbs' [3] work on statistical mechanics, but the first applications of the concepts to the science of information theory should be attributed to Claude Shannon [4] and to A. I. Khinchin [5]. An excellent and comprehensive discussion has been provided by Brillouin [6] and this reference is highly recommended to the reader.

Predictably, mathematical definition of information has little to do with the usability of the information transmitted by a signal channel. It concerns only the range of possible outcomes permitted by a message.

Specifically, information is defined as being proportional to the log of the possible outcomes in a message. If there are J total elements in the message, and P_j is the number of possible outcomes of the jth result, then

$$I = K\sum_j \ell n\ P_j \tag{1}$$

The proportionality constant K is chosen to meet the operational demands of the problem. In statistical mechanics, the proportionality constant includes Boltzmann's constant and the definition of information becomes directly related to the statistical expression of the entropy of a system. In describing the information capacity of a channel, it is useful, especially in the context of using computer systems, to use as the basic unit, the bit. (The transmitting of information can be viewed as making different yes-no selections; hence, the binary system serves well.) Then,

$$I = \ell n_2\ e\sum_j \ell n\ P_j \tag{2}$$

or

$$I = \sum_j \log_2 P_j \tag{3)*}$$

The expression (3) above gives us the amount of information in a message with J characters. Let the message be a 5-letter word, and each letter in the alphabet have an equal probability of being used in each of the characters. (That this is not true in any given word is obvious.) Then,

$$I = \sum_{j=1}^{5} \log 26 = 5 \cdot 4.70 \tag{4}$$

or

$$I = 23.502 \text{ bits} \tag{5}$$

Hence, a 5-letter word which includes any possible alphabetic combinations may be encoded in 24 bits.

Since many minicomputers today are 16 bits, an alternate question which has practical value is: How many letters of a 26-letter alphabet may be encoded in a 16-bit word?

*From this point on, $\log_2 N$ will simply be $\log N$. Ln N will imply the Naperian logarithm.

$$16 = J \cdot \log 26 \tag{6}$$

therefore,

$$J = 3.4 \tag{7}$$

Therefore, a 16-bit word may encode 3 letters when the base set is limited to 26 characters. The complete set of standard teletype characters, both upper and lower case, includes 128 possibilities. A single-character TTY message requires 7 bits to encode; conversely, 2 characters may be encoded in a 16-bit word.

B. Signal Analysis and Generalized Uncertainty

The fundamental purpose of most instrumentation in the chemistry or physics laboratory is to generate a signal and perform or provide for the analysis of that signal in its time or frequency components. Regardless of whether the domain of the signal is wave number, wavelength, mass units, frequency, or time, all domains can be mapped into either a time or frequency domain. Time (t) and frequency (f) are the two basic conjugate spaces we will consider here; they are Fourier transform pairs.

No instrument is capable of providing an exact spectrum analysis:

$$G_0(f) = \int_{-\infty}^{\infty} g_0(t) \exp(-j2\pi ft)\, dt \tag{8}$$

Each experiment must be turned on at some time (t = 0) and must be turned off at some time (t = T). The spectrum analysis performed, in practice, is

$$G(f) = \int_{0}^{T} g_0(t) \exp(-j2\pi ft)\, dt \tag{9}$$

Equations (8) and (9) may be restated by defining a measurement window m(t), which in the simplest and ideal case, has the properties:

$$m(t) = 1 \qquad 0 \le t \le T \tag{10a}$$

$$m(t) = 0 \qquad t > T,\ t < 0 \tag{10b}$$

Note that, in general, m(t) will not be unity over the interval for which it is nonzero, due to instrumental imperfections in the real world. The measurement function in frequency space is the Fourier transform of m(t).

$$M(f) = \frac{\sin 2\pi f(T/2)}{2\pi f} \exp\left(-2\pi j \frac{T}{2}\right)$$

The factor $\exp[-2\pi j f(T/2)]$ results from the definition of m(t) over 0 to T rather than over -T/2 to +T/2. It is a simple phase modulation of the signal and will be ignored in this discussion. The $\sin 2\pi f(T/2)/2\pi f$ factor quite importantly defines our resolution, Δf. The first zero of this function occurs at $f = T^{-1}$; therefore, the full width defined at the zero crossing of an infinitely narrow spectral line (Dirac δ function) is $\Delta f = 1/2T$. Rayleigh defined his resolution criterion similarly--only from an operational viewpoint. He determined that in the diffraction limit of a dispersive spectrometer, if the maximum of one diffraction pattern were superimposed on the first node of its neighbor's, the two lines could be distinguishable by an intensity dip of about 19%. Rayleigh's measurement frequency function was of the form $(\sin x/x)^2$ (and his measurement window was triangular (not square) in its nonzero region).

From the above discussion, we can intuit* an important form of the uncertainty principle, namely, that the product of the observation time T and the desired resolution Δf are greater than some constant:

$$\Delta f T \geq \frac{1}{2} \tag{12}$$

So far, we have discussed spectral resolution but not spectral bandwidth; and the observation time window, but not the time resolution within that window. The spectral bandwidth W and the time resolution also form an uncertainty relation, and this relation has important consequences in the application of computers to real time laboratory experiments.

---

*For a rigorous derivation, see Ref. 7. Our discussion assumes that physical systems are a good approximation of linear time-invariant systems.

Consider one definition of a Dirac δ function in time:

$$\delta(t) = \int_{-\infty}^{\infty} \exp(i2\pi ft)\, df \tag{13}$$

The Dirac time function, or a time impulse, contains all the frequencies in spectral space. If we define a function h(t) as the impulse response of a system, then the Fourier transform of that function describes the frequency response of the function, or the behavior of the system throughout its bandwidth. That the system is causal (no response until the system is turned on) demands that the real and imaginary parts of the frequency response function $H(\omega)$ be related by Hilbert transform pairs [8,9]. The output, in time, of the system to any input is given by

$$s_0(t) = \int_0^t h(t')s_i(t - t')\, dt' \tag{14}$$

The spectral output is given by

$$S_0(f) = H(f)S_i(f) \tag{15}$$

H(f) and h(t) are related by

$$H(f) = \int_0^{\infty} h(t) \exp(-j2\pi ft)\, dt \tag{16}$$

In practice, both h(t) and H(f) are smoothly varying analytic functions. For purpose of example, consider that

$$h(t) = \frac{\sin(2\pi f_1/2\ t)}{2\pi f_1/2} \tag{17}$$

Then

$$H(f) = 1 \qquad 0 \le f \le |f_1| \tag{18a}$$

$$H(f) = 0 \qquad f > |f_1| \tag{18b}$$

The frequency region $f = 0$ to $f = f_1$ is then the bandwidth of the system, and we have a second form of the uncertainty principle [via an argument analogous to the previous one, where Δt is determined by the first zero of h(t)]

$$W\Delta t \geq \frac{1}{2} \tag{19}$$

This brief discussion has shown that the form of the uncertainty principle familiar to quantum theorists may be applied to both the determination of resolution and the determination of bandwidth in a measurement system.

An interesting aside should be made here. So far, no mention has been made of noise or its role in signal analysis. It can be shown that in the absence of noise, a measurement system can achieve infinite resolution. Slepian and Pollack [10] have shown that the set of functions called prolate spheroidal wave functions, which possess interesting orthogonality properties over both the finite time interval 0 to T and the full domain t, can be used to extrapolate g(t) in Eq. (9) to values of t greater than T. In the absence of noise, infinite resolution is available. To the extent that noise is present in the time interval 0 to T, their process in effect trades S/N for resolution. The technique requires that G(f) be band-limited [G(f) = 0 outside of a finite range of f]. In the case of a real (noisy) measurement system, the gain in information in the strict sense is, of course, zero; and the usefulness of the technique is tenuous at best as would be expected from the second law of thermodynamics (see Sec. II.C).

C. Channel Capacity and Shannon's Measure of Information Change

In Sec. II.A, Shannon's definition of information was presented and applied to a simple 5-letter message. If a set of messages of 5 letters were received at a constant rate, the rate of information in the channel might be defined. If the number of messages per second is W, then the information capacity, C, in bits per second is

$$C = WI \tag{20}$$

and for a 10 message/sec rate, our 5-letter message requires a channel capacity of 235.02 bits/sec. A distinction is to be made here. This is the capacity required of the channel. It is not required that each message sent have 23.502 bits, but in order to allow for all the possible outcomes, the channel must be capable of transmitting information at the full WI rate.

Via a rigorous derivation, Shannon has shown that the required capacity of a discrete noiseless channel is given by

$$C = \lim_{T \to \infty} \left[\frac{\log N(T)}{T} \right] \tag{21}$$

where N is the number of allowed signals of duration T.

The definition of the information required to transmit a message has told us nothing about the information actually transmitted by the message. Shannon devised a quantity to handle this problem as well. If the initial state of the system [6] has P_0 equal a priori probabilities, and an amount of information I_1 is required to reduce the number of possibilities to P_1, we have

$I_0 = 0$ $\quad$ P_0 possibilities
$I_1 > 0$ $\quad$ P_1 possibilities

and

$$I_1 = \log \frac{P_0}{P_1} \tag{22}$$

Shannon's measure of the uncertainty before a message is produced is given by

$$S = K \sum_{j=1}^{J} p_j \log p_j \tag{23}$$

where the p_j are the probabilities of the jth outcome. (This is less restrictive than speaking of the total number of outcomes P, because it allows for unequal probabilities.) This function has several unique characteristics:

1. It is zero of any $p_j = 1$ and the others are zero (the state of the system is known).
2. It is a maximum for p_j all equal (the case of highest disorder).
3. It can express a conditional measure based on the p_j being conditional probabilities, i.e.,

$$S(i|j) = \sum_{i,j} p(i|j) \log p(i|j)$$

Shannon proposed that the measure of information be the difference in the S value of the system before the message, and after the message be the measure of the information in the message. He called his measure entropy because of its clear relation to the thermodynamic quantity. Tribus [11] reports that the name was suggested by John von Neuman first, because it had described a similar phenomenon in statistical mechanics, but second, because the subject was sufficiently misunderstood by enough people so that in debate, Shannon would always be at an advantage. Of course!

Brillouin [6] enlarged on the subject and introduced the concept of negentropy which more completely describes information transmitted by bound systems (a bound system refers to a finite physical system). In this case, the total entropy is constant, as required by the second law of thermodynamics. Remembering Eq. (22),

$$I_1 = \log P_0 - \log P_1 = S_0 - S_1 \tag{24}$$

Hence,

$$S_1 = S_0 - I_1 \tag{25}$$

The "bound" information appears as a negative term in the entropy equation for a system. Therefore, the transmission of information results in a <u>decrease in the entropy</u> of the system (Brillouin's term: negentropy).

As we are primarily interested in defining the capacity of a signal channel, this discussion of entropy as a measure of information may seem to be a diversion. But the definition given in Eq. (21) is for a discrete noiseless channel, and as such, is not representative of the "real world." In the real world, there are two sources of statistical fluctuation: one is the signal source itself, and the second is the channel. If the entropy of the input source is S(i), and the entropy of the output channel is S(j), in the noiseless case S(i) = S(j), but in the noisy case

$$S(i,j) = S(i) + S(i|j) = S(j) + S(j|i) \tag{26}$$

Shannon [4] has shown that the channel capacity for a noisy channel is

$$C = \mathrm{Max}[S(i) - S(i|j)] \tag{27}$$

i.e., the channel capacity required is the maximum difference between the uncertainty of the source [S(i)] and the conditional uncertainty S(i|j) of the source (i) given the uncertainty of the channel (j). The latter term Shannon referred to as the equivocation of the channel.

For a continuous noisy channel, where S(y) is the entropy of the continuous source, and S(n) is the entropy of the noise, a similar expression applies

$$C = \mathrm{Max}[S(y) - S(n)] \tag{28}$$

When the noise distribution is white Gaussian, the entropy per second of the received data can be expressed as

$$S(y) = W \log 2\pi e(P + N) \tag{29}$$

where P is the signal power and N is the rms value of the noise. Similarly, the entropy of the noise source is

$$S(N) = W \log 2\pi eN \tag{30}$$

This gives us a very interesting and useful relationship for the channel capacity which covers most laboratory experiments, namely,

$$C = W \log \frac{P + N}{N} \tag{31}$$

or

$$C = W \log (1 + \mathrm{SNR}) \tag{32}$$

Since Eq. (32) requires that the maximum difference in entropies be used, the SNR (signal-to-noise ratio) implies that we evaluate the maximum anticipated SNR.

D. Sampling the Signal: Another Shannon Theorem

In the same reference [6] Shannon proposed what is commonly known as the sampling theorem for band-limited signals. If $g(t)$ is limited to contain no frequencies beyond W, i.e., $G(f)$ is contained in $(\emptyset, W)$, then

$$g(t) = \sum_{-\infty}^{\infty} X_n \frac{\sin \pi(2wt - n)}{\pi(2wt - n)} \tag{33}$$

where

$$X_n = g\left(\frac{n}{2w}\right) \tag{34}$$

The X_n are equally spaced samples of $g(t)$ in the time domain. The sampling distance, Δt, is given by

$$\Delta t = \frac{1}{2w} \tag{35}$$

The sampling frequency is

$$f_s = 2W \tag{36}$$

Shannon's theorem states simply that two samples of signal $f(t)$ per cycle of W yields enough information to reconstruct the signal $f(t)$ for all values of t between the samples, provided that the signal contains no frequencies higher than W. Since this condition holds for any $f_s > 1/2w$, it is easily seen that Eq. (34) is a restatement of the uncertainty expression (19).

The basic parameters required for designing laboratory systems are now available: a definition of channel requirements based on signal bandwidths and signal-to-noise, and a rule for sampling and signal reconstruction. The reader is referred to Shannon [4], Brillouin [6], and Tribus [11], for a more detailed and enriched discussion of these topics.

III. SYSTEM DESIGN AND MACHINE REQUIREMENTS

A. Machine Requirements

It should be noted, first of all, that the author's preference is toward the use of the "minicomputer" for laboratory applications. Modern semiconductor technology has made available almost limitless computer power at very reasonable cost. The choice of minicomputer should be made in terms of features required by the total system, software and service support provided by the manufacturer, and price [12]. Evaluation of the processor features may be subordinated into three categories: instruction repertoire and machine architecture, I/O organization, and interfacing.

1. Instruction Repertoire and Machine Architecture

Internal minicomputer instruction sets deal with several categories of instructions [12]. These are (i) arithmetic and logical instructions, (ii) load and store instructions (memory access instructions), (iii) branch instructions, and (iv) I/O (input/output) instructions.

The first category of instructions normally involves the use of an accumulator (a register internal to the CPU) and another accumulator or memory location. (The second alternative is preferable.) The second category is normally required to transfer individual datum from one memory location to another or to load to or store from accumulators during an arithmetic or logical process. The final category is that which gives a computer its intelligence without which it cannot be really programmable. The branch instruction enables a computer to choose what instruction it is to perform next based on the results of a previous nondeterministic operation, be it arithmetic, logical or data collection from the outside world. In Fig. 1 is a short assembly language program which illustrates these features. It is coded in the assembly language of the Data General NOVA. Text to the right is explanatory comment and not computer instruction. In this trivial program, there are the rudiments of the concepts above. Lines 3 and 4 are load instructions (memory reference); lines 1 and 2 are data or data pointers; lines 6 and 7

```
1.       A:                           ;Label A for the constant 5.
2.       B:  Pointer                  ;Pointer to a data area.
3.  Start:   LDA 0,5                  ;Program start, load accumulator
                                      ;zero with the constant 5.
4.           LDA 1, Pointer @         ;Load the contents of the address
                                      ;pointer to accumulator 1.
5.           Sub 0,1 SZR              ;Subtract 0, 1, skip the next in-
                                      ;struction if the results are zero,
                                      ;skip the next instruction if the
                                      ;results are not zero, execute the
                                      ;next instruction.
6.           JMP LOC1A                ;Other program locations.
7.           JMP LOC1B                ;Other program locations.
```

FIG. 1.

are part of a branch operation which is initiated as part of the arithmetical/logical instruction in line 5.

The ability to address data and to jump to given locations is also an important feature of a minicomputer. A typical 16-bit machine will have four types of addressing ability: absolute addressing, relative addressing, register addressing, and indirect addressing:

1. Absolute addressing refers to the ability of the program to directly access a memory location by its absolute sequential address. This is usually restricted to the first $256_{(10)}$ locations in memory.
2. Relative addressing is the addressing of some memory location within $\pm 128_{(10)}$ locations of the address of the currently executed statement (program counter).
3. Register addressing is addressing a location held in one or more of the accumulators of the CPU. This range is normally 15 bits in a 16-bit machine, or 32,768 locations.
4. Indirect addressing is similar to register addressing, except that instead of a register, the address to be accessed is contained in another memory location, the contents of which are accessed via the absolute, relative, or register method.

Some comments are in order. First, the trivial program in Fig. 1 shows the very fundamental nature of assembly language programming.

This is a drawback of many minicomputers but more will be said about this subject in the next section on operating systems. Second, it is extremely difficult at the user level to compare machine architectures and instruction sets in a meaningful way. The economy of any assembly language program is as dependent on the programmer as on the hardware, and the economy of a higher level program is similarly dependent on both the operating system and the machine hardware.

2. Input/Output Organization

The most important feature of a minicomputer used in the laboratory is the organization of its input/output (I/O) structure. This organization determines with what ease a user may effect:

1. Individual control and status I/O via processor accumulators
2. Generation and handling of interrupts from high priority devices
3. Interrogation of the status of a device
4. Input or output of data streams via data channels or direct memory access (use of the CPU is not required for each datum transferred)

External devices access the processor or memory via a bus. Figure 2 shows a typical bus (I/O bus, Unibus, etc.), as used in a Digilab FTS-14 spectrometer system. Each device is assigned a number which is encoded and decoded at both ends. All transfers to and from the processor are preceded by a signal which has encoded the source or destination device.

For low priority or low data rate transfers, data may be transferred under normal program control. In an environment where there are several real time processes taking place, the processor interrupt structure and the direct memory access feature may be employed. From the programming point of view, the device is set up to begin its transfers (this usually involves at least outputting to the device a starting memory location and a word count), and the program proceeds with some other task. The external device will "steal" from the processor at least one-cycle of CPU time during each transfer (this is normally about one-half of an instruction execution

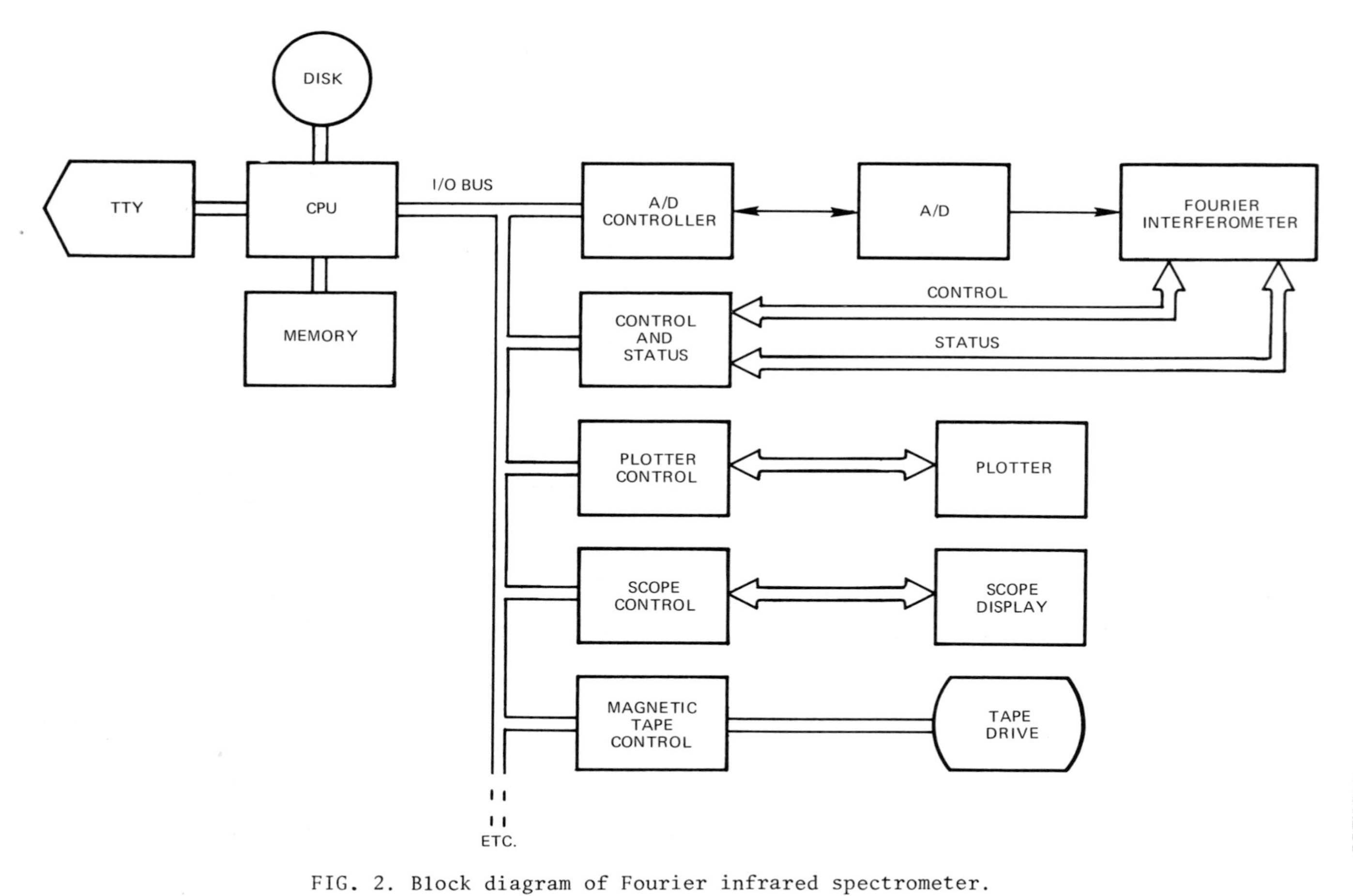

FIG. 2. Block diagram of Fourier infrared spectrometer.

time). When the device has completed its transfers, an interrupt is declared. The interrupt is a hardware signal which causes the program counter to jump to a specific location in memory. This location contains the start, or points to the start of an interrupt handling routine. The first obligation the interrupt service routine has is to preserve the state of the world at that time, i.e., to save all active registers, etc., so that after servicing the interrupt, the program may continue in an orderly fashion. In some computers, this obligation is met via hardware, which is a very, very nice feature!

Interrupts may be nested as with "Do" loops in FORTRAN, and the method of nesting is closely related to the way the hardware distributes priority on the I/O bus. Priority may be assigned via a mask (usually 16 levels are available). Within a given priority level, priority between two devices is determined by their physical proximity to the processor on the bus.

Without a reasonably sophisticated interrupt structure, real-time device handling in the laboratory is virtually impossible.

3. I/O Interfacing

Some brief remarks can be made about I/O interfacing. Each device on the bus requires a controller. This controller must perform the device decoding, and have at least one register. Each computer manufacturer has specific timing requirements for encoding data transfers on the I/O bus. Most manufacturers offer generalized device handlers, which when purchased, put a minimum requirement on the user for further logic design. Alternatively, there are many systems vendors who perform this interfacing service as well.

The evaluation of interfacing methods is subjective, but two rules are offered:

1. All interfacing should use TTL logic wherever possible.
2. Devices on the bus should always be buffered in both directions.

B. System Design

Long-range planning and conservative estimation are the watchwords of the system designer. His basic tools are block diagrams (used to see the "whole picture") and timing diagrams (used to calculate the details of the channel capacity).

Remember that our definition of information and of channel capacity is logarithmic and, therefore, the channel capacity requirements for any device or the ensemble of devices (the I/O bus or the CPU) are additive.

The block diagram for a Fourier infrared spectrometer is shown in Fig. 2. The signal-to-noise ratio of the interferogram in a given scan in the mid-infrared region is often on the order of 32,000:1. The bandwidth of the interferogram is approximately 1250 Hz. The channel capacity required of the sampled signal is then

$$C_I = 2 \cdot 1250 \log (1 + 32{,}000) \tag{37}$$

or

$$C_I = 37{,}500 \text{ bits/sec} \tag{38}$$

Since the system under consideration signal averages the incoming signal, the signal-to-noise ratio builds up in the disk memory. The channel capacity required of the I/O bus is then on the order of

$$C_B = C_I + 2C_S \tag{39}$$

where the C_S is the channel capacity required to transfer a 32-bit dynamic range interferogram between the disk memory and the core memory (once for a read, and once for a write).

$$C_S = W_{DISK} \log (1 + S/N) = 57835 \cdot 32 \tag{40}$$

$$C_S = 1.85 \times 10^6 \text{ Hz}$$

Then C_B is

$$C_B = (3.75 \times 10^4 + 3.70 \times 10^6) \text{ bits/sec} \tag{41}$$

The capacity of the data channel for the NOVA 1200 is 555,555 16-bit words/sec. This is an average bit rate of 8.9×10^6 megabits.

It would seem that the data channel is fast being absorbed by the disk, and that the processor is being saturated, at least from the viewpoint of our watchword concerning conservative estimation. However, this is where the detailed timing diagram is important. Consider Fig. 3.

For each 6.55 sec of signal scan, the I/O bus is used by the disk for only 2.266 sec. During peak usage the I/O bus is used at 40% efficiency. On the average, the usage is at 14% of its capacity. For 65% of the time, it is used at 0.4% of its capacity.

For applications to multiple experiments, programming considerations can lead to almost full utilization of the I/O bus and processor time. Processor time must always be considered, because it is the clock with which the I/O bus reckons. If the data channel steals every other machine cycle for I/O transfers, a program execution time can be cut in half. This leads to the expression "I/O bound." If the processor must perform a series of computations which enable I/O transfers, and the I/O devices are awaiting the resolution of these computations, the process is "compute bound."

CONDITIONS: SCAN TIME: 6.55 SECONDS
ARRAY SIZE: 16,384 SAMPLES
SAMPLING RATE: 2,500 Hz
DISK TRANSFER RATE: 57,000 Hz
DISK BLOCK SIZE: 256 WORDS
DISK SPEED: 3,600 RPM

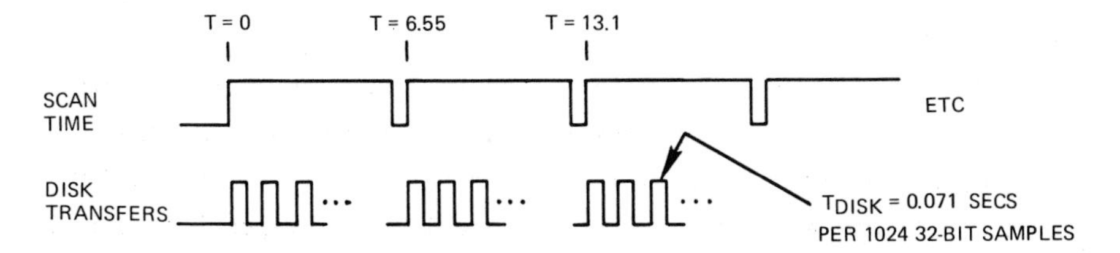

FIG. 3. Timing diagram for NOVA 1200. There are 32 sets of transfers (reads and writes). Each transfer is 0.0708 sec duration. Total time used by the disk is 2.266 sec. Time available for other functions is 4.28 sec.

IV. THE REAL TIME OPERATING SYSTEM

Implementing a real-time laboratory computer system has the potential for being so complicated a task that the difficulties involved so far outweigh the advantages, and that the cost effectiveness of the project might disappear. The complexity and detail involved in assembly language programming can be seen in the trivial example in the previous section. The allocation of processor time and memory space which more fully utilizes the I/O capability in Sec. IV.A is also a formidable task. One approach which relieves some, but not all, of these burdens is the use of an operating system [14-17].

The operating system primarily performs a management function both for the user and the programmer. The resources which the operating system manages are memory, processors, devices, and information. In a secondary role, the operating system also supports the activity of program generation: assemblers, compilers, editors, debuggers, etc. While these are not necessary features of an operating system in general, for application in a real-time laboratory environment, the integration of these library facilities with the operating system is a necessity.

Figure 4 shows the interaction of the operating system in its management role with its subordinate resources from a user viewpoint. The user in this case need not be considered a person, but rather as a function or a task. Users 1 and 2 might be two laboratory experiments started by User 3, but with whom he will have no interaction until the tasks are completed. User 3 might be an experimenter, or a programmer.

Specific features of operating systems are best given by example. Examples here will be restricted to the Data General Real Time Disk Operating System (RDOS) [15]. Further examples of the MIT Multics system may be found in Refs. 14 and 17. Reference 16 provides some insight into the SAGE air defense operating system and the SABRE airlines reservation real time operating system.

The operating system, RDOS, will be described in terms of its

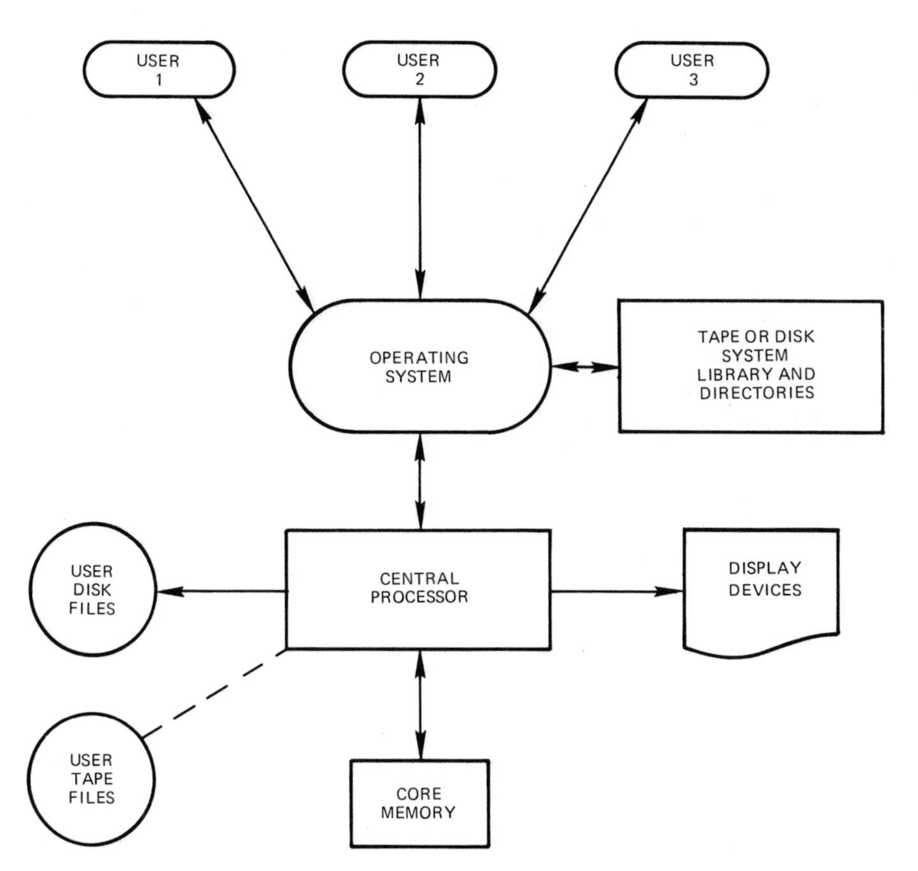

FIG. 4. Diagram of interaction of real time operating system with users.

organization, its operational modes and its methods of communication. Where appropriate, references will be made to laboratory applications. No discussion of system generation will be entertained; the interested reader is referred to Ref. 15.

A. System Organization

From the perspective of the system, the organizational features most directly rule the usage of core memory and disk space. Figure 5 is a schematic outlining these delimiters. The disk directories

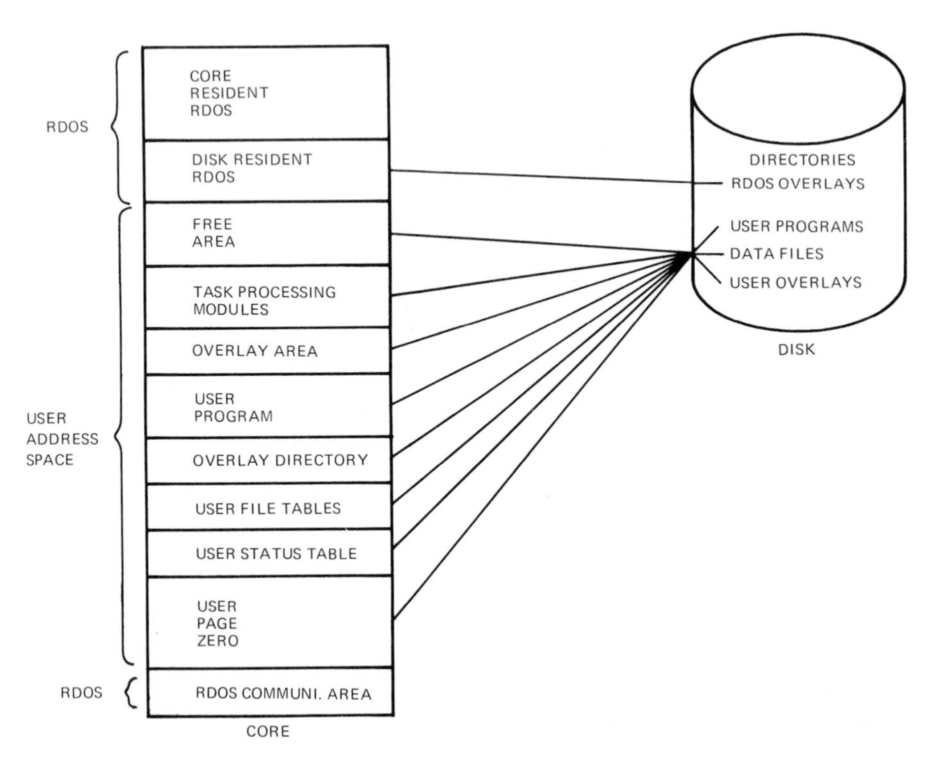

FIG. 5. Diagram of RDOS in terms of core memory and disk space.

allow the system to locate its own operating programs, its overlays, the user's programs, the user program overlays, and the user data files.

The lowest $16_{(8)}$ core locations and a segment of core in the top of memory are reserved for the operating system. The low core area from location $16_{(8)}$ through $377_{(8)}$ is available to the user. A user status table (UST) is required for each operating program level; this table describes the program length, the number of tasks, and the I/O channels which the user program requires. Above the UST is the user file table, one table for each I/O channel used. The last part of the user status table includes an area reserved for task control blocks. These control blocks provide the system with the status information of each task in the queue. (This will be further discussed in the following section.)

All programs are activated or deactivated to one of several states via the operating system. Programs are loaded in and out of core via the system's relocatable loader. The relocatable loader will load relocatably assembled core images into specific memory areas. While the code is relocatable, the loader is not dynamic, and every time a program is activated, it is loaded into the same fixed space within the user area.

B. System Operating Modes

RDOS permits the execution of multiple tasks under a foreground and background environment. Typically, high priority tasks are performed in the foreground, while low priority tasks are performed in the background. The three levels of the system operation are the command line interpreter (CLI) level, foreground/background level, and the task level.

The CLI is the basic means of communication with the system. It is used for executing (i) system support programs: editors, assemblers, compilers, etc.; (ii) system overhead functions: list disk files, disk space, etc.; (iii) creating data or program files: defining names, attributes, etc.; (iv) initiating execution under foreground and/or background operation.

Figure 6 shows the flow of system control from a bootstrap (CLI only) condition through to foreground/background operation. Only a background program may overwrite the CLI directly, although a background program may initiate foreground operation. This is shown in the right half of the flow diagram. The left half indicates that the initiated foreground program must be coexistent with the CLI.

The task level is a sublevel of the foreground/background level, but it is the level at which the true power of a real time operating system unfolds. A task is a logically complete path through an integral program unit. This program unit may be an entire program, a subprogram, or an overlay. The power of the operating system is its ability to handle multiple tasks (not necessarily part of the same

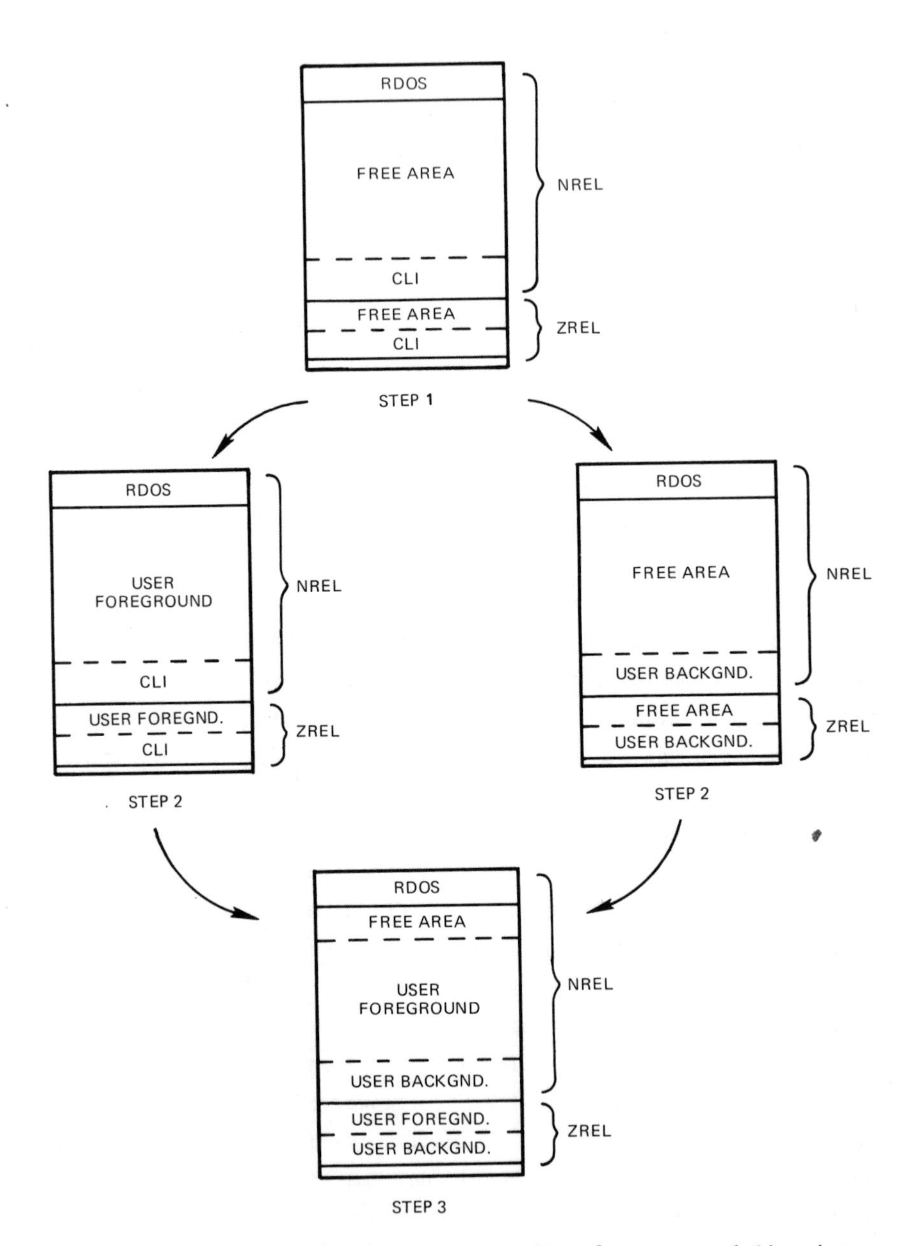

FIG. 6. Diagram of system control flow from command line interpreter to foreground/background level. ZREL = page zero relocatable memory; NREL = normally relocatable memory.

program) in real time. It is at this level that the operating system truly manages the CPU, the memory, and the peripherals of the system. Each task is assigned a priority, which priority may be changed by the system or the user depending on the initiation of other tasks of higher or lower priority.

A simple example in the laboratory might be a three-task environment. Task A is data collection from a laboratory device whose initiation is completely controlled by the system, but once a run is initiated, the data is transmitted from the system according to its own internal clock. Task B is the plotting of previously recorded data onto a digital X/Y plotter. Task C is the analysis of previously acquired laboratory data, which is stored on a disk file.

In the operating system, there are four states which tasks assume: Executing, Ready, Suspended, and Dormant. This is shown in Fig. 7. At initialization, all tasks are dormant. Assume that the user initiates the system into task C, the computation. C is now Executing. He further requests Task B: that the file data be displayed. Until the system Suspends C in order to initiate B, task B is Ready. The user then requests the start of task A. Task B is then Suspended to start A. After A is initiated, B and C are then serviced in some order depending on their priority. Assuming that A (the data collection) would be assigned the highest priority, either B or C might be Suspended whenever the I/O device transmitting from the laboratory needs service. All of these tasks would be executed under the foreground operation of the system. This is an example of a multitask environment. While RDOS can be programmed to give equal time (time-sharing or time-slicing) to each task; this would not be done in this case.

Specifically, while task C might require most of the CPU time, it has the lowest priority. No disaster would occur if the next multiplication is postponed for a short time period. However, task A, the experiment, might use the least of the CPU time, but would have the highest priority. If the system fails to service that task in time, the data might be lost and the experiment might be a failure.

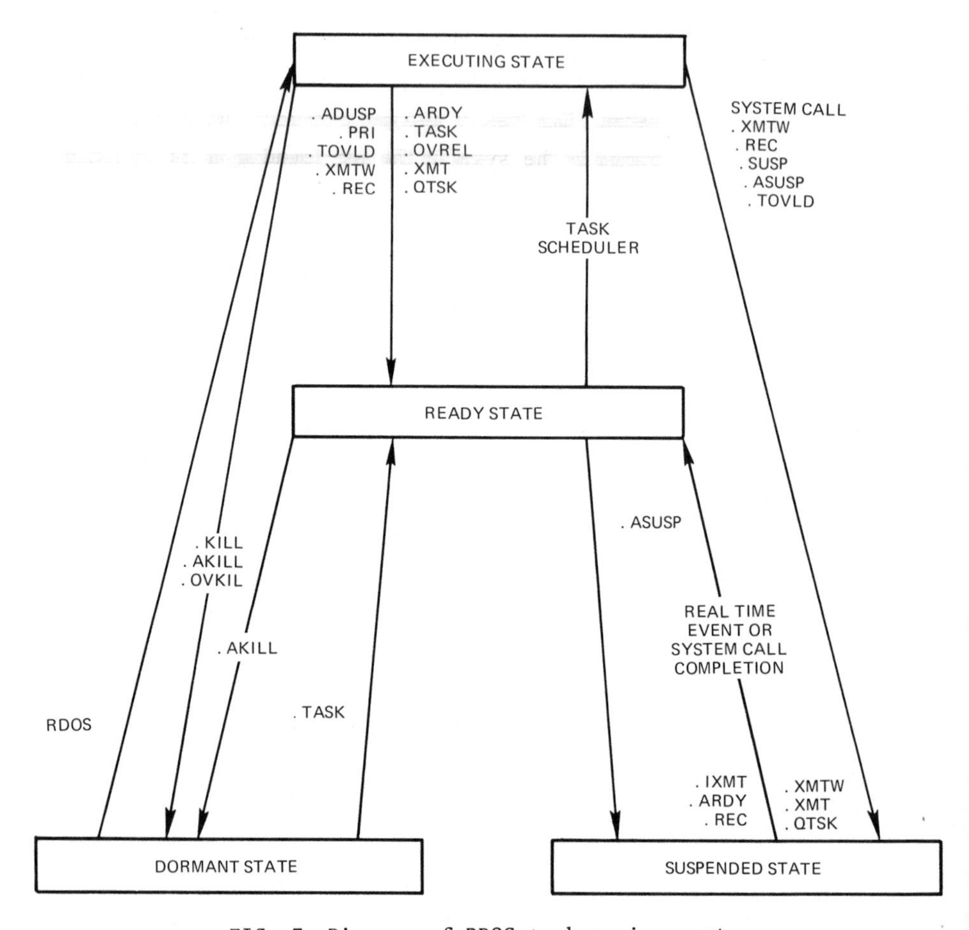

FIG. 7. Diagram of RDOS task environment.

An interesting aside should be made here. All programs run in this multitask environment should be reentrant; this allows a once suspended and now reactivated task to share the same code with other tasks and with itself (in time). RDOS FORTRAN and ALGOL are both reentrant.

C. System Communication and I/O Processing

Two of the most frequently used methods of communication within the system are .SYSTM calls and .TASK calls. .TASK calls are used within the programs to start, interrupt, kill, or change the priority of an executing or nonexecuting task.

.SYSTM calls are 2-word instructions which are used to (i) manage data files in disk and core, and (ii) manage the transmission of I/O data between system or user files and I/O devices. Some examples of .SYSTM calls are listed below:

.SYSTM .RDL (.WRL)n	Read (write) a line to I/O channel n.
.SYSTM .RDB (.WRB)n	Read (write) a block of data to channel n.
.SYSTM .Rename	Rename a file.

The I/O technique above, where operationally distinct amounts of data (line, block, etc.) may be transferred with 2-word instructions is one of the system features which makes the system task more realizable.

Nonstandard I/O devices may be controlled directly by their device codes, or the user may write special device drivers and incorporate them in the operating system. $76_{(8)}$ device codes are available (some are assigned by the system for standard channels) and channel control information such as disk address, core address or word count are passed in accumulators. Use of channel code $77_{(8)}$ enables the actual channel code to be passed in an accumulator for flexible runtime channel selection.

In the previous section (III.B) we discussed an example in which a laboratory device required high peak channel capacity but low average channel capacity. Figure 8 shows a schematic of single-task channel usage versus multitask RDOS channel usage for four devices (card reader, TTY input, disk, and TTY output) and the CPU.

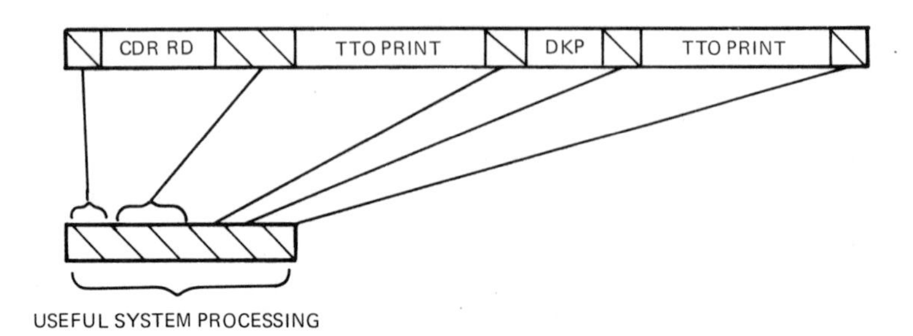

SINGLE TASK OPERATING SYSTEM

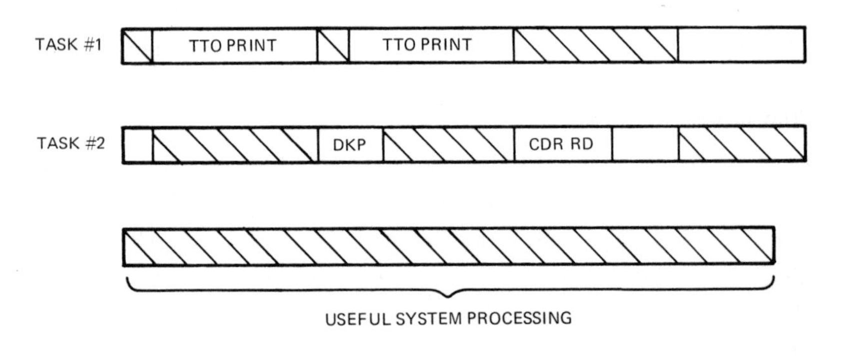

MULTIPLE TASK OPERATING SYSTEM

I/O PROCESSING OR TASK SUSPENSION

USEFUL SYSTEM PROCESSING

FIG. 8. Diagram of single-task and multitask RDOS operating usage.

D. Some Additional Comments on Resource Management in Real Time Laboratory Applications

If several different channels are collecting data at high bandwidths, the tradeoff between the size and number of core buffers as a function of the number of users and their respective required bandwidths becomes somewhat subtle. Goldsmith [18] has derived a set of disk buffering equations which cover these contingencies. His case assumes that not only are the users collecting data, but that they are signal averaging large arrays onto the disk, and that the disk array size exceeds available core.

There are Nu users, and the following definitions apply:

1. Tn is the time for the disk to make one revolution.
2. Ti is the time for the ith A/D to collect sufficient samples to fill a buffer as large as the number of points accessible in a single disk revolution (a user may have an array which is q times larger than this).
3. Ni is the amount of core memory required (in units of disk buffers of one length rotation) by the ith user.

The Goldsmith equation for Nu users,

$$Ni \geq Ti^{-1}\left[Ti + (Nu + 1)Tr + \sum_{j=1}^{J} Nj \frac{2TjTr}{(Tj - 2Tr)}\right] \tag{42}$$

says that the amount of core (in units of disk buffers) is greater than or equal to the product of the reciprocal unit collection time and the sum of the unit collection, the total latency generated by the other users and an aggregate term related to the time required by the other users to both fill their buffers and transfer that data to the disk. In the expression above, there are J + 1 = Nu users; the expression refers to the requirements of the ith user in terms of the other J users.

The solution of these equations is a problem in linear programming. The object is to simultaneously minimize the total of all the Ni (core buffers) for the Nu users.

Goldsmith has also pointed out that in an operating system environment, it is necessary either to take account of possible latency introduced by the operating system (Tr becomes larger) or to prevent the operating system from arbitrarily performing disk transfers during critical collection periods. Both solutions are realizable.

V. CHANNEL REQUIREMENTS FOR INFRARED SPECTROMETERS: AN EXAMPLE

Analytical infrared spectrometers can be separated into two categories: dispersive spectrometers (grating or prism) and nondispersive (e.g., Michaelson/Fourier type). The basic wave number range ($\nu_u - \nu_1$) of interest is from 4000 cm^{-1} to 400 cm^{-1}. Typical resolution ($\Delta\nu$) requirements run from 8 cm^{-1} to 0.1 cm^{-1}. Given a sample, the dependent variable in the experiment is S/N; the independent variables are resolution and scanning or observation time for the total wave number range.

Under defined conditions [19], it can be shown that a Fourier infrared spectrometer possesses a signal-to-noise advantage (Fellgett's advantage) over a dispersive spectrometer of at least $\sqrt{N}$, where N is the number of resolution elements in the spectrum. This advantage is not without price, which price is paid in the capacity of the signal channel handling the data, as we shall see.

Let us consider first the dispersive experiment, in which the desired signal-to-noise ratio at the 0% absorption line in the spectrum is 500/1. Assume that this is achieved at 2 cm^{-1} resolution and that the response time of the detector requires that the observation time over the entire spectrum be 30 min (1800 sec). Further assume a linear scanning speed. Therefore, the 2000 resolution elements are scanned in 1800 sec; the minimum sampling rate is twice for each resolution element, so

$$C = \frac{2 \times 2000}{1800} \log (1 + 500) \text{ bits/sec} \tag{43}$$

or

$$C = 20 \text{ bits/sec} \tag{44}$$

Note that the maximum channel dynamic range required is nine bits, since that is the maximum signal in the spectrum, and the data is sampled directly in the wave number domain.

The Fourier spectrometer [19] on the other hand, collects data in the spatial domain; the spatial coordinate x represents the motion of a moving mirror in the Michaelson interferometer. Since the mirror moves in time $x = vt$ (where v is the mirror velocity), the signal output, called an interferogram (b), can be written either as a function of space or time

$$b = b(x) = b(vt) \tag{45}$$

The interferogram is the sum of many frequencies each linearly related to a spectral wave number:

$$b(x) = \int_0^\infty B(\nu)\ \exp(-2\pi j\nu x)\ dx \tag{46}$$

or

$$b(vt) = v\int_0^\infty B(\nu)\ \exp(-2\pi j\nu vt)\ dt \tag{47}$$

Each spectral wave number then has a corresponding audiofrequency (related by the mirror velocity):

$$f[\mathrm{Hz}] = \nu[\mathrm{cm}^{-1}]V[\mathrm{cm/sec}] \tag{48}$$

The sampling frequency is determined from the maximum spectral frequency ν_u,

$$f_s = 2\nu_u V \tag{49}$$

typically for 4000 cm^{-1}, f_s = 2500 Hz

Assume a nominal spectral dynamic range in a single Fourier spectrometer scan (at 2 cm^{-1} resolution) is 150/1 for a 4-sec measurement. A signal-to-noise ratio of 500/1 is achieved by signal averaging 16 consecutive scans (improving the resultant S/N by 4), for a total measurement time of 64 sec.

The increased sensitivity comes from Fellgett's advantage; but in realizing this advantage, the interferogram must be digitized, not the spectrum. The dynamic range in the interferogram is $\sqrt{N}$

times greater than the dynamic range in the spectrum. Since the spectral dynamic range is 500/1, for a 2000 resolution element spectrum the interferogram dynamic range is 6700/1. The channel capacity requirements are then

$$C = 2\nu_u V \log (1 + 6700/1) \tag{50}$$

or

$$C = 31{,}775 \text{ bits/sec} \tag{51}$$

where the dynamic range of the digitizer must be at least 13 bits.

This example serves to indicate the difference between information and channel capacity. Both the dispersive spectrometer and the Fourier spectrometer transmitted the same mathematical information (the spectrum); the increased sensitivity of the Fourier spectrometer allowed this same information to be transmitted in a much shorter time. This increased rate of information puts a much larger demand on the capacity of the information channel.

REFERENCES

1. *The Last Whole Earth Catalogue*, Portola Institute, Random House, New York, 1971.
2. D. Alcosser, R. Phillips, and H. Wolk, *How to Build a Working Digital Computer*, 1967).
3. W. Gibbs, *Elementary Principles of Statistical Mechanics*, Yale University Press, New Haven, Conn., 1902.
4. C. E. Shannon and M. Weaver, *The Mathematical Theory of Communication*, The University of Illinois Press, Urbana, Ill., 1949.
5. A. I. Khinchin, *The Mathematical Foundations of Statistical Mechanics*, Dover, New York, 1949.
6. L. Brillouin, *Science and Information Theory*, 2nd ed., Academic Press, New York, 1962.
7. D. E. Vakman, *Sophisticated Signals and the Uncertainty Principle in Radar*, Springer-Verlag, New York, 1968.
8. A. Papoulis, *The Fourier Integral and Its Applications*, McGraw-Hill, New York, 1962.
9. R. Bracewell, *The Fourier Transform and Its Applications*, McGraw-Hill, New York, 1965.

10. G. Slepian and J. Pollack, *Bell System Technical Journal*, Vol. 4 (1961).

11. M. Tribus and E. C. McIrvine, Energy and information, *Sci. Am.*, Sept. (1971).

12. *How to Buy a Minicomputer*, Data General Corporation, Southboro, Mass., 1972.

13. D. E. Knuth, *The Art of Computer Programming*, Vols. I, II, Addison-Wesley, Reading, Mass., 1968.

14. J. Donovan, *System Programming*, McGraw-Hill, New York, 1972, Chap. 9.

15. *Introduction to the Real Time Disk Operating System*, Data General Corporation, Southboro, Mass., 1972.

16. J. Martin, *Programming Real Time Computer Systems*, Prentice-Hall, Englewood Cliffs, N. J., 1965.

17. S. Rosen (ed.), *Programming Systems and Languages*, McGraw-Hill, New York, 1967, Chap. 5.

18. M. A. Goldsmith, A Real-Time Time-Sharing System for Analytical Chemistry, Unpublished Thesis, MIT, Cambridge, Mass., June, 1973.

19. P. Griffiths, C. T. Foskett, and R. Curbelo, Rapid scan infrared Fourier spectroscopy, in *Applied Spectroscopy Reviews*, Vol. 6 (E. G. Brame, Jr., ed.), Marcel Dekker, New York, 1973.

Some additional references for those interested in information theory and science are as follows:

R. B. Evans, A Real Proof that Essergy is the Only Consistent Measure of Potential Work, Ph.D. Thesis, University Microfilms, Ann Arbor, Mich., 1969.

D. Gabor, Light and information, in *Progress in Optics*, Vol. 1 (E. Wolf, ed.), North-Holland Publ., Amsterdam, 1961.

M. Tribus, Information theory as the basis for thermostatistics and thermodynamics, *J. Appl. Mech.*, March (1961).

M. Tribus, P. T. Shannon, and R. B. Evans, Why Thermodynamics is a Logical Consequence of Information Theory, *A.I.C.E. J.*, March (1966).

AUTHOR INDEX

Numbers in brackets are reference numbers and indicate that an author's work is referred to although his name is not cited in text. Underlined numbers give the page on which the complete reference is listed.

SUBJECT INDEX